YOUR FIRST 30 DAYS WITH AI

BUSINESS EDITION

PRACTICAL ACTIVITIES TO SAVE TIME, CUT COSTS, AND SUCCEED WITH AI ACROSS YOUR ORGANIZATION

JULIE ANNE EASON

ISBN: 978-1-944602-38-3

Book formatting by Amit Dey
Publishing assistance by Kevin LeBlanc

Publisher's Cataloging-in-Publication
(Provided by Cassidy Cataloguing Services, Inc.)

Names: Eason, Julie Anne, author.

Title: Your first 30 days with AI : 300 practical activities to save time, cut costs, and succeed with AI / Julie Anne Eason.

Other titles: Your first thirty days with artificial intelligence

Description: Business edition. | [Garner, North Carolina] : [Thanet House Publishing], [2025]

Identifiers: ISBN: 978-1-944602-38-3

Subjects: LCSH: Artificial intelligence. | Business enterprises--Technological innovations. | BISAC: COMPUTERS / Artificial Intelligence / Generative AI. | BUSINESS & ECONOMICS / Careers / Career Advancement & Professional Development. | BUSINESS & ECONOMICS / Business Communication / General.

Classification: LCC: HD30.2 .E27 2025 | DDC: 658.0563--dc23

DEDICATION

To the coming generations,
may your future be bright.

CONTENTS

INTRODUCTION

Welcome to the future. If you've heard about AI's potential in marketing but aren't sure where to start, you're in the right place. This guide is designed to walk you through practical, hands-on projects, helping you integrate AI into your marketing strategy with confidence, without feeling overwhelmed. By the end, you'll be using AI to save time, spark creativity, and get more done in less time.

When I first tried using AI, I'd heard all the buzz and decided I wanted to see if it lived up to the hype. Everyone said, "Just mess around and you'll figure it out." Well, let me tell you, that's terrible advice. I went in without a plan, got bad results, and quickly decided AI was a complete waste of time.

It's important to go into AI with a purpose. You need to have something specific you're working on or a result you want to achieve, like creating a social media content calendar or drafting an email campaign. That's the difference between getting frustrated and getting real value out of the technology. And that's exactly why I'm writing this book—to help you learn how to use AI one step at a time.

We'll start with the basics and gradually build your confidence, but we'll do it through active projects. You'll learn by doing, with practical tasks that serve real purposes for your business or organization. No more hoping for good results through luck. This guide will show you how to get there deliberately, with a focus on consistency and deliberate practice.

There's a common belief that AI is going to "steal" people's jobs, but the truth is that AI is more likely to enhance those jobs. Some jobs will undoubtedly be phased out, that happens with every leap in technology. But when you are skilled in using AI, you're more valuable to the workforce of the future. For people who take the time to learn these skills, AI is going to make their jobs easier by handling repetitive tasks, freeing them up to focus on more strategic and creative work.

By embracing AI, you can get more done in less time, allowing you to focus on what really matters.

I'm Julie Anne Eason, and if there's one thing you should know about me, it's that I've spent my entire career helping people get their voices heard. I've ghostwritten books for CEOs, coached entrepreneurs, and worked with some of the most creative minds in business to help them turn their ideas into compelling stories that inspire, educate, and sell.

I didn't set out to become an AI consultant. I approached AI with skepticism at first. I've built my 25-year career on storytelling and marketing, crafting messages that feel human and sound like the brand or person I'm writing for. That's not an easy thing to do. So, when I first heard about AI tools that could supposedly do the writing for you, I was sure it was all just hype. But as I started digging into the technology, I realized that AI is an incredibly powerful tool. It isn't here to replace writers, marketers, or business leaders; it's here to enhance what we do. And one thing's for sure, everyone is going to need these skills just to do their everyday jobs.

I'm not a tech guru. I'm a business owner like you, trying to make the most of the tools we have. If you can write an email, you can use AI. And that's why I'm writing this book. I want to take the confusion out of AI and show you exactly how it can work for your business, just like it works for mine.

Why Marketing?

We're using marketing projects for a simple reason: every business needs marketing, and every business needs to create content. It doesn't matter if you're a small business owner, part of a nonprofit, or working in a larger organization, content creation is a cornerstone of reaching any audience.

And when it comes to AI, creating content is one of the easiest and most practical ways to get started. It's something you can see immediate results from, and it's a skill you'll use over and over again in your business. Marketing also allows for easy experimentation and quick feedback, which is ideal for learning AI. So, we're beginning with marketing because it's universal, and it's the perfect entry point to learning how to integrate AI into your daily work.

Marketing involves reaching your audience with the right message at the right time, but creating content, doing research, and staying ahead of trends can be time-consuming and repetitive. AI will help by generating ideas, writing content, analyzing data, and even building entire marketing campaigns—all much faster than you can do manually.

30 Days, Step by Step

This guide is broken down into four weeks of projects, each designed to help you build your AI skills and directly benefit your business:

- **Week 1:** You'll learn the basics of AI prompting and set up your AI Resource Center, where you'll store prompts, brand guidelines, and other key outputs for future use.
- **Week 2:** We'll dive into advanced techniques. If you're a Pro user, you'll learn how to build a Custom GPT to fit your brand and audience. If you're using the free version of ChatGPT, you'll practice advanced prompting to create consistent results across sessions.
- **Week 3:** You'll see how AI can help with research and strategy. You'll use AI to analyze competitors, generate content ideas, and even perform basic data analysis to improve your marketing campaigns.
- **Week 4:** You'll pull everything together by using AI to create a complete lead generation funnel. This includes developing a lead magnet, writing landing page copy, creating an email nurture sequence, and drafting promotional content.

Each week builds on the last, so by the end of the guide, you'll be comfortable using AI for a variety of marketing tasks.

NOTE: Throughout this guide, you'll see sample prompts and exercises to help you learn each concept and skill. You're welcome to follow these samples as written to understand how they work, or you can substitute your own company's details to make each exercise more practical for your needs. That way, you'll be accomplishing real work while you learn!

Go at Your Own Pace

You don't need to follow the 30-day schedule exactly. Feel free to move faster or slower depending on your needs and schedule. Some tasks will only take a few minutes, while others might require more exploration. Take your time to practice and get comfortable, there's no rush. And if you feel like you need to revisit earlier activities, that's perfectly fine.

I also encourage you to follow your curiosity. That's where the fun is. If you're unsure whether AI can handle a task, just ask it! If the answer is yes, it will give you a detailed step-by-step guide on how to do it. If the answer is no, it can give you alternatives to try. The more you experiment with it, the better you'll understand its capabilities. Asking questions and trying new things is how you'll discover what AI can truly do for you.

I like to think of AI as a five-year-old child. It knows everything, but it doesn't understand anything. It can write about almost any topic, and it can answer almost any question you throw at it, but here's the catch: like a five-year-old, it needs clear instructions. You have to give it specific parameters, context, and guidelines to make sure it delivers what you actually want. Without those directions, it might wander off into strange or irrelevant territory. Learning to give AI the right instructions—that's the skill of prompting. When you can do that well, AI becomes an incredibly powerful tool.

If you're not particularly "techy," don't worry, you'll be fine. If you can write an email, you can learn to use AI. The tools may seem complex at first, but they're designed to be user-friendly. This guide will break down each task step-by-step so you can follow along easily, no matter your skill level. We'll take it one step at a time. AI is here to help you, not complicate things. You don't need advanced technical skills, just a willingness to explore and learn.

For any bumps along the way, refer to the troubleshooting guide in the appendix to get back on track quickly.

Understanding the AI Landscape

The AI world is evolving fast, with new tools and updates being released constantly. While this might seem overwhelming, the skills you'll learn in this guide will remain valuable, no matter what changes come. Even though there are hundreds of AI platforms available, the core concept—working with a large language model (LLM) like ChatGPT—stays the same. The way you interact with AI through clear, structured prompts is consistent across platforms. These skills will also help you understand and navigate future AI updates. The sooner you start learning, the more prepared you'll be.

Many of the AI tools out there are just repackaged versions of the same underlying technology. For instance, Microsoft's Copilot is a white labeled version of ChatGPT. It may look different or have a unique interface, but it's fundamentally the same technology. Once you've learned how to prompt and work with a tool like ChatGPT, you'll be able to apply those skills to nearly any generative AI platform. The technology might change, but the foundational skills you develop here will always be relevant.

Staying Safe: Privacy and Data Security in AI

Before we dive into the practical steps, let's address a crucial topic that's on many business leaders' minds—privacy and data security. When working with AI tools like ChatGPT, it's natural to wonder how your data will be handled and how to keep your information secure. The good news? With a few simple precautions, you can confidently leverage AI without compromising privacy.

Here are a few best practices to keep in mind as you start:

1. **Avoid Sharing Confidential Information:** While AI tools are incredibly powerful, remember that they aren't immune to data vulnerabilities. Avoid sharing sensitive details such as customer data, financial information, or proprietary strategies within any AI platform.
2. **Use Summaries and Anonymized Data:** If you need to analyze or generate ideas around sensitive business scenarios, consider summarizing or anonymizing details. This way, the AI can still provide valuable insights without accessing sensitive specifics.
3. **Consider Enterprise Solutions and Private Setups:** For larger-scale AI implementations, enterprise-grade solutions often come with enhanced security features, including data encryption, access controls, and audit trails. Additionally, some organizations are opting for private language models (LLMs) hosted on their own servers. These custom setups keep all data secure in-house, allowing for a fully controlled AI environment that aligns with your company's security and compliance requirements.
4. **Review the Platform's Privacy Policy:** Each AI platform has its own approach to data storage and usage. Familiarize yourself with their policies to ensure they align with your privacy standards. For instance, OpenAI, the creator of ChatGPT, has privacy guidelines in place and allows users to disable chat history to prevent their data from being used for further model training.

By following these practices, you can make the most of AI's benefits while safeguarding your data. Rest assured that throughout this book, the focus is on practical, secure applications of AI that support—not compromise—your brand's integrity.

What to Expect Over the Next 30 Days

This guide isn't just about learning new skills. It's about creating marketing materials you can immediately apply to your business. Whether it's writing blog posts, drafting email campaigns, or analyzing data, every task you complete will have practical, real-world value. You'll also be saving everything in your AI Resource Center, where you can refine or reuse your work whenever needed. I also suggest keeping a journal or notes on your progress to help you track your learning and maintain motivation.

As with any new tool, you may not get perfect results from AI right away. Sometimes, the AI might misunderstand your prompts or give you outputs that need refining. This is normal, and part of the process is learning how to adjust your prompts and get better results over time. Don't be discouraged if it takes a little trial and error. Each interaction helps you improve your skills and build your confidence.

Free vs. Pro: Understanding Your Options

Throughout the guide, we'll use ChatGPT to complete tasks. If you're a paid user with a ChatGPT Plus subscription, you'll have access to GPT-4, which allows you to build a Custom GPT tailored to your brand. This means you won't need to write detailed prompts for every task. You can train the AI once and use it repeatedly. (Don't worry, I'll explain all this in Week 2.)

If you're using the free version of ChatGPT, you'll work with GPT-3.5. While you won't be able to create a Custom GPT, you can still achieve strong results by writing detailed prompts. You'll learn how to craft effective prompts to guide the AI toward the outputs you need, even without the extra features.

We strongly encourage you to consider upgrading to ChatGPT Plus if you want to explore more advanced capabilities. The subscription is affordable, and you can cancel at any time. That said, the free version is still a powerful tool, and you'll be able to complete all the tasks in this guide. Keep in mind that AI tools evolve quickly, and free users may have access to more features by the time you're reading this.

Let's Get Started

You're beginning a journey that will transform the way you approach marketing. Whether you're brand new to AI or you've dabbled in it before, this guide is designed to meet you where you are and give you the tools to move forward with confidence. Remember, AI isn't about replacing what you do, it's about enhancing it. It's a powerful tool that will help you work smarter, not harder, and give you more time to focus on what really matters in your business.

As you go through these activities, keep an open mind and don't be afraid to experiment. AI is here to support you, and the more you interact with it, the more you'll see how much it can offer. You'll be building practical, real-world marketing assets as you go, so by the end of this guide, you'll have something tangible to show for your efforts—not just a new skill set, but results you can apply immediately.

So, take a deep breath, trust the process, and get ready to discover how AI can make your life easier. You've got this, and I'm here to guide you every step of the way. Let's dive in and start building something great together!

PRE-ASSESSMENT: WHERE ARE YOU STARTING?

Before diving into Week 1, take a moment to reflect on your starting point with AI in marketing. Use the following questions to assess your confidence and goals. You don't need to do anything with your answers yet. By this time next month, you'll be able to see how far you've come. Feel free to write your answers directly in the book.

On a scale of 1 to 10, how confident do you feel about using AI in marketing right now?

1-3: I've heard of AI, but I don't know where to start.

4-6: I've tried AI tools, but I don't feel confident using them regularly.

7-9: I use AI occasionally but haven't fully integrated it into my workflow.

10: I'm confident and use AI regularly in my marketing.

Your rating: ________

What do you want to learn most about AI?

What marketing tasks do you hope AI can help with?

Reminder: Don't rush through. Take the time to really reflect on where you are right now. If you're stuck or unsure at any point in the book, it's okay to go back and revisit earlier activities. Revisiting earlier activities is also a great way to reinforce learning and build confidence.

CREATING AND USING AN AI RESOURCE CENTER

An AI Resource Center is a centralized space where you can save and organize all of your best prompts, successful outputs, and insights about using ChatGPT (or any other AI tool). Think of it as your personal AI playbook—a guide you can quickly reference as you work through this book and beyond. You'll streamline your workflow, reduce time spent re-creating prompts, and ensure that your AI outputs remain consistent in tone, style, and quality over time.

When you first start working with AI, it's easy to assume you'll remember the prompts that work well. But as you experiment with different tasks and refine your approaches, it becomes surprisingly difficult to keep track of everything mentally. An AI Resource Center helps you:

1. **Save Time:** Once you've crafted a great prompt or template, you won't need to reinvent the wheel every time you need similar content. Instead, you can quickly locate a proven prompt and reuse or adapt it.
2. **Ensure Consistency:** For professionals, maintaining a consistent tone, style, and structure across different content is essential. Your Resource Center allows you to document your brand's tone and style guidelines, making it easier to apply them consistently.
3. **Track What Works (and What Doesn't):** Your Resource Center can help you keep track of insights, patterns, and strategies that improve the quality of your AI outputs. By noting what worked well and what didn't, you'll continually refine your approach and get better results over time.
4. **Adapt to Different Audiences and Formats:** As you experiment with AI-generated content, you'll likely find that certain prompts work better for specific audiences or content types. By categorizing and saving these in your Resource Center, you can quickly find the right prompt for the right situation.

To keep your Resource Center organized and easy to navigate, consider setting it up with these key sections and adding to it as needed:

1. **Prompts & Templates:**

   This is where you'll save successful prompts, categorized by content type (e.g., social media, emails, blog posts, product descriptions). For each saved prompt, include a short description of when and why it worked well, so you can easily identify the best one to use for future tasks. Over time, this section will become a valuable library of templates you can adapt to various situations.

2. **Notes & Insights:**

   As you work with ChatGPT, jot down any observations about what worked well and what didn't. For instance, you might note that asking the AI to "act as a brand strategist" produced a better result than simply asking it to "write a product description." This section can also include any troubleshooting tips you discover, common pitfalls to avoid, and any patterns you notice about how the AI responds to specific types of prompts.

3. **Audience Profiles:**

   If you're creating content for multiple audiences (e.g., small business owners, B2B clients, consumers), use this section to document each audience's key characteristics. Include details on preferred tone, style, common concerns, and key interests for each group. This will help you tailor prompts for different audiences and ensure that the AI's output resonates with them.

4. **Tone & Style Guides:**

   Every brand or organization has its own voice and style, and consistency is key. Use this section to document specific tones and styles that work best for your brand. For example, if you want your content to be "friendly yet professional" or "engaging and inspirational," write down examples of prompts and outputs that achieved that tone. This is especially helpful if multiple people are using the Resource Center, as it keeps everyone aligned on brand voice.

5. **Content Examples:**

   Sometimes, the AI produces outputs that are particularly effective, creative, or insightful. Save these standout examples here for future reference. You can repurpose these as inspiration or templates for new projects, saving time and ensuring high-quality results.

6. **Revision Strategies:**

   No AI response is perfect on the first try. In this section, document techniques you find helpful for refining AI outputs, such as how to ask the AI to adjust tone, length, or format. For example, you might note that saying "make it more concise" tends to work well when you need shorter responses, or that "rephrase for clarity" helps simplify complex information.

How to Set Up Your AI Resource Center

This can be as simple or sophisticated as you'd like. Here are some options for organizing all your notes:

- **Digital Note-Taking Tools:** Tools like Notion, Evernote, or OneNote allow you to create dedicated sections, embed links, and add tags for easy searching. These platforms are ideal if you prefer a digital, searchable format that's accessible from any device.
- **Spreadsheets:** If you're comfortable with spreadsheets, create columns for each section (e.g., prompt, description, content type, notes) and use tabs to separate different types of content. Google Sheets or Excel can work well for this approach, offering both simplicity and functionality.
- **Text Document:** A Word document or Google Doc works well if you're looking for a straightforward, low-maintenance option. You can create headings for each section, use bullet points for notes, and add links to specific chat URLs if you want to reference past conversations.
- **Physical Notebook:** If you prefer an offline approach, a dedicated notebook can work too. Use dividers or color-coded tabs for each section, and write out prompts, notes, and examples by hand. This approach may take a bit more effort to keep organized, but some people find it helpful for focusing and avoiding digital distractions.

Tips for Maintaining Your AI Resource Center

1. **Update Regularly:** AI tools are constantly evolving, and so are your needs. Make it a habit to update your Resource Center regularly with new prompts, templates, and insights.
2. **Save Successful Outputs Promptly:** Whenever you get an AI response that's particularly effective, save it immediately in your Content Examples section. It's easy

to assume you'll remember it later, but having it stored will save you time down the road.

3. **Refine Over Time:** As you gather more data, feel free to re-categorize prompts, add new sections, or delete things that aren't helpful. Your Resource Center should be a flexible, evolving resource that adapts to your needs.
4. **Back Up Your Center:** If you're using a digital tool, consider backing it up periodically. This way, you won't lose any valuable prompts or notes if something goes wrong with your device or platform.

As you work through this book and experiment with different prompts, add successful examples and useful insights along the way. Over time, this resource will become an invaluable tool, saving you time and helping you get the most out of AI.

WEEK 1 OVERVIEW

GETTING STARTED WITH AI TOOLS AND BASIC PROMPTING

Welcome to Week 1! This week is all about laying a strong foundation with AI tools and learning the essential skill of prompting—how to "speak" to AI effectively to get the results you want. Don't worry if AI feels unfamiliar at first. We'll start with the basics and build your confidence through hands-on practice.

What You'll Learn This Week

This week focuses on three core areas:

1. **Understanding AI Tools and Setup**
   - How to navigate ChatGPT's interface effectively
   - Setting up your AI Resource Center for organized, efficient work
   - Managing different chat threads and saving important conversations
2. **Mastering Basic Prompting**
   - The Who/What/How framework for clear communication with AI
   - How to structure requests for better responses
   - Techniques for refining and improving AI outputs
3. **Creating Your First AI-Generated Content**
   - Writing social media posts and engaging captions
   - Drafting professional emails and communications
   - Building basic blog posts and FAQs
   - Understanding when and how to refine AI's output

Building Your Foundation

Think of this week as learning a new language—the language of communicating with AI. Just like learning any language, you'll start with basic phrases (simple prompts) and gradually build up to more complex conversations (sophisticated prompting techniques).

The skills you learn this week will serve as building blocks for everything else you'll do with AI. Take your time with each concept and practice until you feel comfortable. It's normal to experiment and adjust as you learn what works best.

Remember, you don't need any technical background to succeed. If you can write an email, you can learn to prompt AI effectively. Focus on understanding the basic principles, and don't worry about getting everything perfect right away. We'll build your skills step by step through practical, hands-on exercises.

Let's get started!

DAY 1

Set Up Your AI Tools and Familiarize Yourself with the ChatGPT Interface

Goal:

Sign up for a ChatGPT account (either free or Plus), familiarize yourself with the ChatGPT interface, and set up an organized AI Resource Center. By the end of today, you'll be comfortable using the prompt box, navigating chat threads, renaming them, and saving chat URLs for future reference.

Activities for Day 1

1. Sign up for ChatGPT (Free or Plus Account)

Go to the ChatGPT website and create your account.

Free Account: This gives you access to GPT-3.5, which can handle most tasks and will work for the activities in this guide.

Plus Account: For a very low fee, you can access GPT-4, which provides better quality outputs and advanced features like creating Custom GPTs (which we'll cover in Week 2). You can upgrade or cancel anytime.

Choose the account that works best for you and get signed up.

2. Familiarize Yourself with the Prompt Box

The prompt box (or context window) is where all the action happens. This is where you'll type your instructions for the AI. You can enter a simple question, task, or detailed instructions.

Try these prompts: Type "What is the importance of AI in marketing?" and press Enter. Watch how the AI generates its response. Next, ask: "Give me three tips for increasing engagement on social media."

...Activity: Now, ask the AI a market-related data question, such as "What percentage of small ...usinesses fail in the first year?" The AI may provide an answer, but it's important to verify it.

Check for accuracy: After the AI gives you a response, type "Did you make that up? Please provide sources with citation links." Evaluate how the AI responds and whether it provides verifiable sources.

3. Switch Between Chat Threads and Rename Them

ChatGPT saves different conversations in separate threads, so you can keep your prompts and outputs organized. You can rename each thread for easier reference later.

Start a new chat by clicking "New Chat" on the sidebar. Ask a different question (e.g., "What's a good marketing strategy for small businesses?"). To rename the chat, click on the name of the chat in the left sidebar and change it to something specific, like "Marketing Strategy Chat." Switch between the two threads by clicking on each in the sidebar. Notice how renaming helps you quickly identify the topic of each conversation.

4. Upload Documents into the Prompt Box

ChatGPT allows you to upload documents for analysis or content creation. We'll explore this more in Week 2, but for now, familiarize yourself with how this works.

Click on the paperclip icon in the prompt box to upload a simple document (like a PDF or Word file). You can ask the AI to summarize the document, but don't worry too much if it's not perfect—we'll cover this in more detail later.

5. Set Up Your AI Resource Center and Save Chat URLs

Open whatever tool you're using to serve as your AI Resource Center. This is where you'll save prompts, templates, and AI-generated outputs as you work through the guide.

Sections to Create: Prompts & Templates, Writing & Output Examples, Brand Voice & Tone Guidelines, Notes & Ideas. You'll be able to add more sections as needed.

Save your chats for future reference: Chat threads can disappear after a while with regular use, so it's a good idea to save the URL for important threads.

To save your chat URLs, click on the three dots next to the thread in the sidebar and copy the URL for the chat. Paste this URL into your AI Resource Center under the relevant section, so you can quickly access it later.

6. Understand the Different Models (For Plus Users)

If you're using a Plus account, you'll have access to different AI models like GPT-4 and GPT-3.5. There may be more models or different names available over time, but the core functionality remains the same: higher-tier models generally offer more accuracy and better responses, especially for complex tasks.

How to Choose Models: When starting a new chat, you can switch between models in the drop-down menu at the top left of the chat window.

If you have a Plus account, start a new chat thread and try switching between available models to see the difference in response quality.

7. Practice Refining Prompts for Better Results

Sometimes, you might need to rephrase or add detail to a prompt to get more relevant responses. Learning how to adjust prompts helps you get better results without having to start from scratch.

Type, "Explain the benefits of content marketing." If the response feels too generic, add more specifics, like "Explain the benefits of content marketing for a small nonprofit organization with a limited budget." Notice how adding more context or specific details improves the response. Experiment with a few different variations and observe what works best for your needs.

8. Edit a Previous Prompt Using the Pencil Icon

Sometimes, after seeing the AI's response, you may realize you could get a better answer by tweaking your original prompt. ChatGPT allows you to edit previous prompts using the pencil icon, which can save time and keep your chat organized.

When to Use the Edit Feature: Use the edit feature when you want to make a minor adjustment to your question or clarify it slightly. This is helpful if the AI's response was close to what you wanted but missed the mark in some way. Editing a prompt keeps your chat focused on a single thread, which can make it easier to follow the conversation.

When to Start a New Prompt: If your new question is different enough that it changes the topic or requires a completely new response, it's usually better to type a fresh prompt. Starting a new prompt lets you explore different ideas without altering the context of the original question.

Type, "List some benefits of AI for marketing." Once the AI responds, click the pencil icon next to your prompt and edit it to say, "List three specific benefits of AI for marketing in nonprofits."

...nter to see how the response changes. Practice using the edit feature to refine prompts ...d get more accurate answers. Then, type a brand-new prompt, like, "What are some challenges nonprofits face when implementing AI?" and see how this keeps the chat focused on a different topic.

> **Note:** Occasionally, you may notice that ChatGPT doesn't fully answer a prompt or seems to get "stuck." When this happens, you'll see a "Regenerate Response" button appear, which you can click to prompt the AI to try again. If the entire chat freezes or shows an error message, try refreshing the page. This often resolves the issue and allows you to continue where you left off. If refreshing doesn't work, you can start a new chat. Don't worry—technical glitches are normal, and these quick fixes can help you stay on track.

9. Prompt the AI to Challenge You or Ask Clarifying Questions

ChatGPT is designed to respond directly to the prompts it's given, which means it generally takes your instructions at face value. But sometimes, it can be helpful to encourage the AI to push back a bit, ask clarifying questions, or challenge your ideas. This can lead to more thoughtful, well-rounded answers—especially when brainstorming or exploring complex topics.

How to Encourage the AI to Challenge You: To get the AI to ask clarifying questions, try adding a request like "Please ask any questions if something isn't clear" or "Challenge my assumptions if you think there's a better approach." This tells ChatGPT to respond in a way that's slightly more dynamic, rather than simply delivering a straightforward answer.

Type, "I'm considering writing a guide on digital marketing strategies for small business owners, but I'm not sure what topics to focus on. Can you help me think through the main points and ask any clarifying questions that might improve the structure?" Notice how the AI responds differently than if you simply asked, "What are some good topics for a digital marketing guide?" By encouraging it to ask questions, you may get a response that dives deeper, helping you refine your ideas.

When to Use This Technique: Asking the AI to challenge your assumptions or ask questions is useful during planning, brainstorming, or any time you're trying to dig deeper into a subject. It can help uncover blind spots or generate ideas you hadn't considered.

Encouraging the AI to interact this way can make your sessions feel more collaborative and insightful—almost like having a brainstorming partner on hand.

Common Mistakes to Avoid

1. Skipping the Resource Center Setup

Mistake: Neglecting to create an AI Resource Center right away.

Better: Set up your Resource Center on Day 1 to keep your prompts, templates, and brand notes organized for easy future access.

2. Failing to Name and Organize Threads

Mistake: Using default names for chat threads, making it hard to find specific sessions later.

Better: Rename each thread by topic (e.g., "Social Media Practice") for efficient organization and reference.

3. Not Exploring the Interface Fully

Mistake: Skipping over the full functionality of the ChatGPT interface, such as editing prompts or using different models.

Better: Take time to explore the interface, especially the features for renaming threads, editing prompts, and switching models if you're a Plus user.

4. Forgetting to Save Successful Prompts

Mistake: Assuming you'll remember good prompts, then struggling to recreate them later.

Better: Save successful prompts in your AI Resource Center immediately for reuse.

5. Not Testing Different Models (for Plus Users)

Mistake: Sticking to one model without experimenting with others, even if additional models are available in your plan.

Better: Try out GPT-4 and other available models to see which one provides the best results for different tasks.

6. Neglecting to Verify AI Outputs

Mistake: Assuming that all AI-generated answers are accurate without verification.

Better: Double-check data or facts, especially if using AI outputs for business purposes. Prompt the AI for citations or sources if relevant.

...ion for Day 1

...hich model did you use today? (Free users: GPT-3.5, Plus users: GPT-4 or other available models)

What did you notice about how the AI responds to your inputs?

How comfortable do you feel navigating the ChatGPT interface?

Did you come across any answers that didn't seem quite right? How can you double-check AI outputs?

What do you hope to accomplish by using AI in your marketing?

DAY 2

Introduction to Basic Prompting with the Who/What/How Framework

Goal:

Learn how to effectively communicate with the AI by mastering basic prompting using the **Who/What/How framework**. By the end of today, you'll understand how adding clear context improves the quality of AI responses and helps you get better results.

Activities for Day 2

1. Learn the Who/What/How Framework

The Who/What/How framework helps you guide the AI to produce the right output for your needs. It's made up of three key parts:

Who: Define who the AI is and who the audience is. Tell the AI what role it should assume (e.g., "You are a copywriting expert" or "Act as a content strategist for small businesses"). Then define the audience for the content (e.g., "You are writing for small business owners who are trying to improve their marketing").

What: Specify the task or content you need (e.g., "Write a social media post about increasing website traffic").

How: Provide additional context or specific instructions for tone, style, or length (e.g., "The tone should be friendly and casual").

2. Experiment: Prompting Without and With the Framework

Part 1: Try a simple prompt without context. Type: "Write a blog post introduction about eco-friendly packaging." Read the response. Does it match your expectations? What tone and audience is the AI assuming?

…ow use the full Who/What/How framework.* Type: "You are a content marketing …t [who the AI is] writing for eco-conscious consumers who are interested in reducing …aste [who the audience is]. Write a blog post introduction about the benefits of using eco-friendly packaging [what the task is]. The tone should be professional and informative [how to deliver the content]." Notice how the response changes when you give the AI more context and instructions.

Reflect on this experiment. How did the outputs differ? Which one felt more useful or accurate?

____

____

____

____

3. Second Experiment: Prompting Without and With the Framework

Part 1: Try a very basic prompt without context. Type: "Write an email about our tote bags." Read the response. Does it sound right for your audience and purpose?

Part 2: Now use the full Who/What/How framework. Type: "Act as a copywriting expert [who the AI is]. Write a promotional email for potential customers interested in sustainable products [who the audience is], introducing a new line of eco-friendly tote bags [what the task is]. The tone should be persuasive and friendly [how to deliver the content]." Compare how the output changes with more specific guidance.

Reflect on this experiment. How did the quality of the outputs change after using the framework? Which prompt gave you a better result? Write down your thoughts.

____

____

____

____

4. Refine the Outputs

If the AI's response still isn't quite right, try refining the blog post introduction or email. For example:

Ask the AI to adjust the tone: "Rewrite the blog post introduction with a more casual tone for a general audience."

Ask the AI to shorten the email: "Condense the email into three sentences for a quick promotional message."

Change the format: Ask the AI to present the content as a list instead of a paragraph. Ask it to provide the information in table format.

5. Prompt Something You Need for Your Business or Organization

Now it's time to put the framework into practice for something you already have to do for your organization. Think about tasks you regularly handle, like writing an email, drafting a social media post, or creating a meeting agenda.

Use the Who/What/How framework to prompt the AI for something specific.

Who: Define who the AI is (e.g., "You are a content strategist for a nonprofit") and who your audience is.

What: Specify the task (e.g., "Write a social media post inviting people to our next fundraising event").

How: Give instructions on tone, style, or length (e.g., "The tone should be friendly and encouraging").

Save the output and refine it if necessary. Make sure to add the prompt and its result to your AI Resource Center.

on Conversational Tone and Prompting Style:

When using ChatGPT, some people like to type prompts in a conversational tone—adding "please" and "thank you" or even joking around with the AI. This is perfectly fine and can make the process feel more comfortable, especially when you're just getting started. You can personalize your prompts however you like; the AI will respond just as effectively.

That said, it's important to remember that ChatGPT is not a human—it's a tool designed to generate text based on patterns in data. While it can be helpful, friendly, and even humorous, it doesn't actually "understand" or "feel" anything. This can sometimes be easy to forget because of how naturally it communicates. Feel free to use whatever style makes you comfortable, but keep in mind that clear, direct instructions will usually get you the best results.

6. Use AI to Improve Existing Content

In addition to creating content from scratch, AI can be a powerful tool for refining and enhancing things you've already written. In this activity, you'll copy and paste a piece of your own writing into ChatGPT and ask for suggestions to make it stronger, clearer, or more engaging. This will help you see how AI can act as an editing assistant, giving you ideas for improvement that you might not have considered on your own.

Step 1: Choose a Piece of Existing Content

Pick a short piece of writing you've already created. This could be an email draft, a social media post, a blog introduction, or even a product description. Aim for something around 3-5 sentences so it's manageable for this exercise.

Step 2: Paste Your Content into ChatGPT

Copy your content and paste it into the ChatGPT prompt box. Add a request for feedback or improvement ideas. For example, you could type:

"I wrote this introduction for a blog post. Can you suggest ways to make it more engaging and informative?"

Or:

"Here's a draft of a social media post I created. Can you suggest edits to make it more concise and compelling?"

Step 3: Review the AI's Suggestions

ChatGPT will respond with suggestions or possible edits to improve your content. Look closely at its suggestions and note which ones you find useful. Does it make your message clearer? Does it suggest more engaging language? Think about how the AI's recommendations align with your goals for this piece.

Step 4: Experiment with Specific Requests

If the initial suggestions aren't quite what you're looking for, try adding more specific instructions to get targeted feedback. For example:

- "Could you rephrase this in a more conversational tone?"
- "Make this sound more persuasive."
- "Help me shorten this to one sentence."

Step 5: Save Your Improved Content

Once you're satisfied with the AI's suggestions, save the revised version in your **Content Examples** section of the AI Resource Center. This will give you a "before and after" example to reference in the future when you need inspiration for editing similar content.

7. Experiment with Missing Elements in the Who/What/How Framework

This activity helps you understand the importance of each element in the framework by intentionally leaving one element out to see how it affects the AI's response.

1. **Who and What, but no How:** Type, "You are a sales expert writing for small business owners. Write a LinkedIn post on the benefits of effective sales follow-up." Notice the tone and format that the AI uses without any specific instructions on style or tone.
2. **What and How, but no Who:** Type, "Write a blog post conclusion about the importance of time management. The tone should be encouraging and motivational." Observe whether the AI assumes an audience or tone that's different from what you had in mind.
3. **Who and How, but no What:** Type, "You are an email marketing specialist creating content for startup founders. Use a friendly, engaging tone." See if the lack of a clear task leads to a more generic response or if the AI makes assumptions about what you want.

After each prompt, notice how the missing element influenced the response.

Framework to Prompt for Different Formats

hat you're comfortable with basic prompting, practice using the Who/What/How mework to request different types of content formats. This will show you how versatile the framework can be, and how it can adapt to various content needs.

1. **Social Media Post:** Type, "You are a digital marketing expert [Who the AI is] writing for freelance designers who are looking to build their brand [Who the audience is]. Write a social media post explaining why a strong personal brand can attract more clients [What the task is]. The tone should be friendly and inspiring [How to deliver the content]."
2. **Listicle:** Type, "You are a productivity coach [Who the AI is] writing for busy professionals [Who the audience is]. Write a list of five quick tips for staying focused during the workday [What the task is]. Use a concise and practical tone [How to deliver the content]."
3. **FAQ Section:** Type, "You are a customer support specialist [Who the AI is] creating an FAQ section for new users of a project management app [Who the audience is]. List three common questions and answers about getting started with the app [What the task is]. The tone should be clear and helpful [How to deliver the content]."

Save the best outputs in your **Content Examples** section for future reference, especially if you frequently create content in these formats.

9. Reverse-Engineer a Prompt

Find a piece of content you like—whether it's an email, social media post, blog excerpt, or marketing blurb—and try to reverse-engineer the Who/What/How framework that could have generated it. This activity helps you think critically about how prompts translate into real-world content.

1. **Choose an Example:** Take a piece of marketing content that resonates with you. For example, a promotional email from a brand you like.
2. **Break It Down:** Identify the likely Who, What, and How elements behind it. Ask yourself:
   - Who was the AI likely instructed to be? (e.g., "You are an experienced brand copywriter.")

- What was the task? (e.g., "Write a promotional email about a new product launch.")
- How did the tone or style come across? (e.g., "The tone should be energetic and inviting.")

3. **Recreate the Prompt:** Write out a prompt using the Who/What/How framework based on your analysis, and then try running it through ChatGPT to see how closely the AI's response matches the original content.

This activity will give you a better sense of how to reverse-engineer prompts, which is a useful skill when you see content you'd like to emulate.

10. Practice Iterative Prompting

Sometimes, the first response you get from the AI might be close but not exactly what you're looking for. Iterative prompting is the process of refining the AI's response step by step, rather than trying to get the perfect result in one go. This activity will help you understand how to work interactively with the AI to shape your content.

1. **Start with a Basic Prompt:** Type, "You are a marketing expert. Write a short introduction to a guide on increasing email open rates."
2. **Review the Output and Identify Gaps:** Read through the response and note any elements that could be improved. Maybe the AI missed specific tips, or perhaps the language is too general. Decide what could make the response more relevant to your needs.
3. **Refine the Prompt:** Based on your observations, add a more specific instruction to get a revised version. For example:
   - "Focus on strategies that small business owners can use."
   - "Mention why email open rates are important for engagement and sales."
   - "Use a conversational tone that encourages readers to take action."
4. **Repeat as Needed:** Continue refining the prompt in small steps, each time making the AI's response closer to your ideal output. For instance, you could say, "Make the introduction even shorter and more direct," or "Add a statistic about average open rates for context."
5. **Save the Final Version:** Once you're satisfied, save the final prompt and output in your **Prompts & Templates** section. This will give you an example of how to use iterative prompting effectively for future tasks.

..ompting is a powerful skill that allows you to get the most out of the AI by work- ..aboratively, rather than expecting perfect results in a single try. Practice this tech- ..e whenever you need to fine-tune content for more specific results.

Common Mistakes to Avoid

1. Not Using the Full Who/What/How Framework

Mistake: Using only partial context in prompts, resulting in vague or off-target responses.

Better: Make sure to include all three elements—Who, What, and How—to guide the AI effectively. This will help you get outputs that are specific to your needs and audience.

2. Skipping Prompt Refinement

Mistake: Accepting the first AI response, even if it's not quite right.

Better: Take time to refine prompts. Adjust tone, format, or add more details to help the AI better align its output with your goals.

3. Focusing Too Much on Technical Precision

Mistake: Worrying excessively about technical wording or jargon in your prompt.

Better: Use straightforward language. The Who/What/How framework is meant to make prompting simple, so avoid overcomplicating instructions.

4. Overlooking Audience Context

Mistake: Not defining the audience clearly, leading to generic content.

Better: Include details about the audience's needs or preferences, such as "small business owners" or "eco-conscious consumers," to create more relevant and tailored content.

5. Ignoring Tone and Style Specifications

Mistake: Leaving out style or tone in prompts, resulting in outputs that don't align with your brand voice.

Better: Specify the desired tone and style, such as "friendly and engaging" or "professional and informative," to ensure consistent brand representation.

6. Not Practicing with Different Content Types

Mistake: Sticking to one type of prompt (e.g., only blog posts), which limits versatility.

Better: Use this practice session to experiment with various formats, like social media posts, emails, or listicles, to understand how the AI handles each.

Review the prompts you've used so far today. Do they clearly specify Who, What, and How? Is the audience defined, and does the tone match your brand? Experiment with revising at least one prompt to include additional context or a clearer style, and see how the AI's response changes. Refine as needed to align with your brand's goals.

or Day 2

d the AI's response differ when you used the full Who/What/How framework?

Which prompt gave you the most useful output, and why?

How comfortable do you feel using the Who/What/How framework to create prompts?

What challenges did you face in getting the AI to deliver the right output?

How do you plan to implement what you've learned today?

DAY 2

DAY 3

Writing Your First Social Media Post with AI

Goal:

Use the **Who/What/How framework** to create your first social media post with AI. By the end of today, you'll be able to prompt the AI to create engaging and on-brand social media content, adapt it for specific platforms, generate multiple posts to support your overall content strategy, and develop basic brand guidelines for consistency.

Activities for Day 3:

1. Start a New Thread and Name It

Open a new thread in ChatGPT and name it **"Social Media Practice"**. This will help you keep track of all the prompts and outputs for today's activities. Start a new thread and name it every day going forward for easy reference.

2. Define the Purpose of Your Post

Start by thinking about the specific purpose of your social media post. Is it to inform, promote, engage, or inspire? Defining the goal will help you shape your prompt and keep your content focused. Common purposes include:

- **Inform:** Share news, updates, or helpful tips.
- **Promote:** Highlight a product, event, or service.
- **Engage:** Ask a question or encourage audience interaction.

Write down the purpose of today's post:

---

3. Create a Prompt Using the Who/What/How Framework

Using the Who/What/How framework, create a prompt that aligns with your purpose and audience. Be specific about who the AI is, who the audience is, what the task is, and how the content should be delivered. Use the sample prompt below to guide you. Replace the placeholder content with your own information.

Sample Prompt:

"You are a social media manager **[who the AI is]** for a small business that sells eco-friendly products **[who the audience is]**. Write an engaging social media post about our upcoming sale on sustainable tote bags **[what the task is]**. The tone should be upbeat and inviting, with a call to action to visit our website **[how to deliver the content]**."

4. Develop Brand Guidelines Using AI

Creating consistent brand guidelines is essential for maintaining a strong presence online. Use the following prompt to help the AI ask you the right questions, so it can compile a customized brand guideline document for you. Replace the placeholder information with your own.

Sample Prompt:

"You are an expert marketer and social media strategist. I run an eco-friendly home goods company online. I want to develop a consistent social media presence. Please ask me the necessary questions to come up with brand guidelines for my company. Once I've answered all your questions, please compile a brand guideline document for me."

Once the AI provides a list of questions, answer them thoroughly to give it the information it needs to create brand guidelines. When finished, the AI will compile a brand guideline document for you, covering essentials like tone, style, target audience, and content themes.

Review the output and refine it by telling the AI what you need changed. When you're happy with it, save this document in your **AI Resource Center** under **Brand Voice & Tone Guidelines** for future reference.

5. Generate Social Media Content: If You Have a Content Calendar or Content Pillars

If you already have content pillars or a content calendar, you can upload these to the AI and prompt it to generate a week of social media posts based on those themes.

Upload your information by clicking on the paperclip icon in the prompt box.

Sample Prompt: "Based on the content pillars in the attached document, create a week of social media posts. Include different tones and calls-to-action where appropriate."

Review the posts and save any that meet your standards.

6. Generate Social Media Content: If You Don't Have Content Pillars

No content pillars? No problem. You can use the Who/What information to ask the AI to suggest relevant content pillars that align with your goals and audience.

Sample Prompt: "You are a social media strategist for a small business selling eco-friendly products. Suggest 3-5 content pillars that would resonate with an audience interested in sustainability. Explain why each pillar would be valuable for this audience."

Review the suggested pillars and choose a few to use as the basis for a week's worth of posts. Then return to activity five to generate the actual content.

7. Build a Social Media Campaign with a Specific Purpose

Let's design a social media campaign for a specific purpose, such as an anniversary sale or holiday special offer. This campaign will include multiple posts with different angles to promote the same event or offer.

Sample Prompt: "Act as a social media strategist for a home goods company celebrating its 5th anniversary with a sale. Create a 5-post social media campaign, including an announcement, countdown, product spotlight, customer testimonial, and final call to action. Each post should be unique and engaging, with a consistent tone and a call to action."

Alternate Sample Prompt for Holiday Offer: "You are a social media manager for a business offering a special holiday discount. Write a 3-post campaign promoting our holiday sale, including a holiday greeting, a post highlighting the discount, and a reminder on the last day of the offer. Each post should be festive and encourage people to shop."

Review the posts and make adjustments to ensure they align with your brand and objectives. Save the full campaign in your AI Resource Center for future reference.

8. Optional Practice: Storytelling Post

Try creating a storytelling-based post to build connection and engagement with your audience. This type of content can share a behind-the-scenes story, customer experience, or even the mission behind your brand.

Sample Prompt: "Write a social media post that tells the story of how our eco-friendly tote bags are made, focusing on sustainable practices. Make it engaging and inspirational, with a call to action to learn more."

Review the output to ensure it aligns with your brand voice. Save the best results for future use.

9. Refine the Posts to Fit Specific Platforms

Social media platforms have different tones and formats, so refine the outputs to better match each platform's style. Try adapting the previous posts for platforms like LinkedIn, Instagram, or X (formerly Twitter).

Sample Prompt:

- "Edit this post for LinkedIn with a professional tone."
- "Condense this for X (formerly Twitter), and add relevant hashtags."
- "Rewrite for Instagram and add an emoji."

10. Save and Organize Your Best Outputs in Your AI Resource Center

Save the best-performing prompts under the Prompts & Templates section. Over time, you'll develop a reliable library of social media templates.

Copy and paste your successful prompts and their refined outputs into the Resource Center.

Common Mistakes to Avoid

1. Unclear Purpose of the Social Media Post

Mistake: Jumping into social media creation without a defined purpose, which leads to unfocused or generic posts.

Better: Start by clarifying the goal of each post, such as to inform, promote, engage, or inspire. A clear purpose guides the tone and structure, making posts more engaging and effective.

2. Forgetting to Define the Audience and Tone

Mistake: Not specifying who the audience is or what tone to use, resulting in content that may not resonate.

Better: Specify the audience and desired tone for each post, whether it's "friendly and conversational" or "professional and informative." This step helps the AI generate content that's tailored and relevant to your followers.

3. Ignoring Platform-Specific Adjustments

Mistake: Using the same content across all platforms without adapting it to each platform's norms.

Better: Refine posts for each platform by adjusting the style, length, or format. For instance, LinkedIn may call for a more formal tone, while Instagram allows for a more casual, visual approach.

4. Overloading the Post with Too Many Messages

Mistake: Trying to convey too many points in one post, which can dilute the message and confuse readers.

Better: Focus on one main point or theme per post. Simplify the content and include a single, clear CTA to encourage engagement.

5. Neglecting to Include a Call to Action

Mistake: Leaving out a CTA, which misses the opportunity to guide audience engagement.

Better: Include a specific CTA in each post, whether it's to "comment your thoughts," "learn more," or "visit our website." A strong CTA drives interaction and helps meet your post's goals.

6. Not Saving Effective Prompts for Future Use

Mistake: Failing to save high-performing prompts, making it harder to maintain a consistent social media voice over time.

Better: Save successful prompts in your AI Resource Center. You'll save time and achieve consistency by reusing and adapting prompts that work well for your brand.

Review the social media posts you've created today. Do they have a clear purpose and a focused message? Is there a defined CTA? Revisit one of your posts to add or refine a CTA, and check that the tone aligns with your audience and platform. Refine as needed to strengthen engagement and consistency.

Reflection for Day 3

What was the purpose of your social media post today, and how well did the AI help you achieve it?

How did the different angles and tones affect the quality of the social media post?

Which platform did you find most challenging to adapt the post for, and why?

How comfortable do you feel using the Who/What/How framework to guide the AI in creating social media content?

What other types of social media content would you like to experiment with in the future?

DAY 4

Writing Promotional Emails

Goal:

Use the Who/What/How framework to create a persuasive promotional email. Today's focus is on prompting the AI to generate emails that capture your brand voice and connect with your audience. By the end of today, you'll know how to craft and refine basic promotional emails that are ready to use or adapt for your business needs.

Activities for Day 4

1. Start a New Thread and Name It

Open a new thread in ChatGPT and name it **"Promotional Email Practice."** Starting a new thread daily helps keep prompts and outputs organized for easy reference.

2. Define the Purpose of Your Email

Think about the specific goal of today's email. Is it to announce a sale, introduce a new product, or re-engage past customers? Define the objective clearly, as this will shape your prompt and focus the AI's response.

Write down the purpose of today's email:

3. Create a Prompt Using the Who/What/How Framework

Using the Who/What/How framework, create a prompt that aligns with your email's purpose. Be specific about who the AI is, who the audience is, what the task is, and how the content should be delivered. Remember you may replace the placeholder details with your own.

Sample Prompt: "You are an email marketing expert **[who the AI is]** writing for eco-conscious online shoppers **[who the audience is]**. Write a promotional email announcing a 20% discount on sustainable home goods **[what the task is]**. The tone should be friendly and persuasive, with a clear call to action to shop the sale **[how to deliver the content]**."

4. Refine the Email to Highlight Key Selling Points

Once the AI generates a draft, you may want to enhance the content by emphasizing certain benefits or features. Or if the first draft doesn't highlight specific selling points well, ask for refinements. The AI never gets tired of rewriting, so ask for as many changes as you need.

Sample Prompt: "Rewrite this email to focus more on the benefits of our sustainable materials and eco-friendly practices." Read the result to see if it aligns with your brand and audience.

5. Experiment with Subject Lines and Calls-to-Action

To make your email more engaging, ask the AI to suggest alternative subject lines or calls-to-action that are attention-grabbing and relevant to your message.

Sample Prompt: "Suggest three catchy subject lines for this email that emphasize the 20% discount."

Once you have subject lines, ask the AI to write alternative calls-to-action, such as "Shop Now" or "Don't Miss Out."

6. Create a Post-Purchase Thank-You Email

A well-crafted thank-you email is a great way to reinforce a positive experience after a purchase.

Sample Prompt: "Write a post-purchase thank-you email for a customer who just bought from our eco-friendly home goods store. The tone should be warm and appreciative, with a reminder of our commitment to sustainability."

If desired, ask the AI to include information about care instructions for the purchased items, links to your store, or an invitation to follow your brand on social media.

7. Follow-Up Request for Review/Testimonial Email

Encouraging reviews and testimonials from customers can help build trust and credibility.

Sample Prompt: "Write a follow-up email requesting a review from a customer who recently purchased eco-friendly home goods. The tone should be friendly and inviting, with a link to leave a review."

You can ask the AI to generate a variation if you prefer the review request to sound more casual or if you'd like it to offer an incentive, like a small discount on the next purchase.

8. Create a Month of Weekly Email Newsletters on a Specific Topic

For ongoing engagement, generate a series of email newsletters to be sent weekly over a month, each focusing on a related theme.

Sample Prompt: "You are an email marketing expert for a brand that sells eco-friendly home products. Create a month of weekly email newsletters focused on 'Sustainable Living Tips.' Each email should be informative, include a relevant tip, and have a friendly, engaging tone."

You can prompt the AI to adjust length, tone, or focus for each email as needed. If the emails look good, save them in your Resource Center for future reference or customization.

9. Testing for Personalization Tags

Adding personalization to emails can make them feel more tailored to each customer and increase engagement.

Sample Prompt: "Rewrite this email to include a greeting with the customer's first name and a discount code placeholder. Make sure it still feels friendly and natural."

Ask the AI to rearrange the personalizations to see if they're more effective. After personalizing the email, save it in your AI Resource Center for future use or as a template for future campaigns.

10. Create a Drip Campaign

For readers interested in automated email marketing, practice creating a drip campaign—a series of emails that gets triggered by customer actions like signing up, purchasing, or abandoning a cart.

Sample Prompt: "Create a 3-email drip campaign for new customers who sign up for our newsletter. The first email should welcome them, the second should introduce our eco-friendly products, and the third should include a special offer to encourage their first purchase."

11. Experiment with Urgency Techniques

Use urgency in emails to encourage immediate action, such as by mentioning a limited-time offer or a sale ending soon.

Sample Prompt: "Rewrite this email to include a sense of urgency by mentioning that the sale ends in 48 hours."

Ask the AI to try a different urgency technique. Once you're happy with the emails, save it in your Resource Center for future use when running time-sensitive campaigns.

12. Write an Email Sequence for Upsells or Cross-Sells

Try creating a follow-up email that suggests complementary or upgraded products to customers after a recent purchase.

Sample Prompt: "Write an email to send one week after a customer purchases a tote bag, suggesting related eco-friendly products they might like."

If the email fits well with your products and audience, save it in your AI Resource Center as a future template.

13. Analyze Email Engagement

If you have access to recent email analytics, such as open rates, click-through rates, or bounce rates, upload these metrics to the AI, along with the appropriate emails, and ask for suggestions to improve engagement. The AI can help interpret these data points to provide relevant strategies.

Click the paperclip icon in the prompt box and upload a simple document or screenshot containing your current analytics.

Sample Prompt: "Review this data from our latest email campaigns and suggest three strategies to improve open rates and click-through rates. Focus on subject line impact, call to action placement, and personalization."

Common Mistakes to Avoid

1. Vague Email Purpose or Objective

Mistake: Sending an email without a clear purpose, resulting in unfocused messaging that doesn't drive action.

Better: Define the specific goal of each email—whether it's to promote a sale, announce a new product, or re-engage past customers. A clear objective makes the email more impactful and aligns with audience expectations.

2. Weak Subject Line

Mistake: Using a generic or unengaging subject line that fails to capture attention.

Better: Spend extra time crafting a compelling subject line that speaks to the email's purpose and entices readers to open it. Consider A/B testing subject lines to see what resonates most with your audience.

3. Overloading the Email with Information

Mistake: Trying to include too much information in one email, which can overwhelm or confuse readers.

Better: Focus on one main message or offer per email. Keep the content concise and easy to scan, and prioritize the most important information at the top.

4. Weak or Missing Call to Action

Mistake: Leaving out a clear CTA or using a vague one, leading to low engagement.

Better: Use a specific, action-oriented CTA that encourages readers to take the next step, whether that's "Shop Now," "Learn More," or "Sign Up Today." A strong CTA drives engagement and helps meet your email's objective.

5. Ignoring Brand Tone and Voice

Mistake: Writing emails that don't align with your brand's tone, resulting in an inconsistent experience for readers.

Better: Keep the brand tone consistent across all emails, whether it's friendly and casual or professional and informative. Use language that reflects your brand personality and reinforces familiarity with your readers.

6. Forgetting to Test and Refine

Mistake: Skipping the testing phase, which can lead to missed opportunities for optimizing email performance.

Better: Test different elements, such as subject lines, CTAs, and layouts, to find what resonates best. Review analytics for each email to inform improvements for future campaigns.

Review the promotional emails you created today. Is each email's purpose clear, and does it align with the subject line and CTA? Check that the tone is consistent with your brand and that the content is focused and easy to scan. Refine as needed to ensure clarity and impact.

Reflection for Day 4

What was the purpose of your promotional email today, and how well did the AI help you achieve it?

How did refining the content affect the email's appeal and clarity?

Which subject line or call to action felt most effective for your audience, and why?

How comfortable do you feel using the Who/What/How framework to guide the AI in creating promotional emails?

What other types of emails would you like to experiment with using AI?

DAY 5

Writing a Complete Blog Post

Goal:

Learn how to use AI to brainstorm, outline, and write a complete blog post, including search engine optimization (SEO). By the end of today, you'll have practiced prompting the AI to generate content that aligns with your brand voice and goals, and you'll know how to refine the post for maximum readability and engagement.

Activities for Day 5

1. Start a New Thread and Name It

Open a new thread in ChatGPT and name it **"Blog Post Writing Practice".** Starting a new thread helps keep prompts and outputs organized for easy reference.

2. Upload Sample Blog Posts for Style Matching (If Available)

If you already have a blog style and tone you like, you can upload a few sample blog posts to help the AI mimic that style. This is especially helpful for maintaining brand consistency.

Click on the paperclip icon in the prompt box to upload your sample blog posts and use this prompt: "Analyze these blog posts to understand my brand voice. Give me an overview of the writing style and tone. Write the next blog post in a similar style."

If you don't have a blog style established, make sure to specify the tone and voice you want the writing to reflect (e.g., "conversational," "professional," or "light and informative"). You can also have the AI mimic the voice of a notable person, blog, or brand.

Save the AI analysis of your brand voice in your Resource Center. You will use it regularly in your prompting.

3. Brainstorm Blog Post Topics

Think about the types of content your audience would find valuable. Use the AI to generate topic ideas that align with your brand's focus, your audience's needs, or seasonal trends.

Sample Prompt: "You are a content strategist **[who the AI is]** brainstorming blog post ideas for eco-conscious readers interested in sustainable living **[who the audience is]**. Suggest five blog post topics related to eco-friendly home habits that would be informative and engaging **[what the task is]**."

If the suggestions resonate, select one topic for today's blog post and save the other ideas for future content.

4. Outline the Blog Post

A clear outline will give structure to the blog post and guide the flow of information. Ask the AI to create a brief outline, breaking the topic into sections.

Sample Prompt: "Outline a blog post on 'The Benefits of Eco-Friendly Cleaning Products.' The outline should include an introduction, actionable tips, and a conclusion with a CTA."

Review the AI's outline, and if needed, ask it to add or adjust sections to better fit the topic or intended length.

5. Write a Compelling Hook

The introduction should capture attention and set the stage for the post. Using the Who/What/How framework, prompt the AI to create an engaging intro hook.

Sample Prompt: "You are a content writer **[who the AI is]** creating a blog post for eco-conscious readers **[who the audience is]**. Write a headline hook for this post on the benefits of eco-friendly cleaning products **[what the task is]**. The tone should be friendly and informative, with a hook to draw readers in **[how to deliver the content]**."

Ask it to provide five alternative hooks to see if it can come up with a more compelling headline.

6. Develop Each Section of the Post

Using the outline as a guide, ask the AI to generate content for each section. Focus on making each section informative and engaging.

Sample Prompt: "Write the blog post. Be sure to include relevant details and examples."

Review the output, and if needed, prompt the AI to refine specific points or add more examples to enhance the section.

7. Add a Conclusion and Call to Action

The conclusion should summarize the key points and include a call to action, such as encouraging readers to make eco-friendly choices or explore related products.

Sample Prompt: "Write a conclusion for this blog post that summarizes the benefits of eco-friendly cleaning products and encourages readers to try sustainable options at home."

If you'd like a stronger call to action, ask the AI to add a line inviting readers to explore your brand's eco-friendly products.

8. Optimize the Blog Post for SEO

SEO optimization can help your post reach a wider audience. Ask the AI to suggest relevant keywords to include naturally throughout the post.

Sample Prompt: "Suggest SEO keywords for a blog post on eco-friendly cleaning products."

After selecting keywords, prompt the AI to integrate them naturally into the content. You can also ask it to revise the title to be more SEO-friendly.

9. Add Supporting Data and Statistics

Adding data or statistics can make the post more credible. Ask the AI to include statistics with source citations and links.

Sample Prompt: "Include a statistic about the environmental impact of traditional cleaning products in the introduction, with a source citation and link."

After the AI provides a source, be sure to double-check the link to confirm it's valid and relevant.

10. Experiment with Tone Variations

If you're unsure about the tone, ask the AI to rewrite parts of the post in different tones to see which best fits your brand.

Sample Prompt: "Rewrite the introduction with a more casual and friendly tone for a general audience."

Experiment with a few tone options for the introduction or conclusion until you find the right match for your audience.

11. Create a Blog Series Introduction

If you'd like to build a blog series on a related theme, prompt the AI to generate an introduction that sets up a series. This approach can help build reader interest and engagement over multiple posts.

Sample Prompt: "Write an introduction for a blog series on sustainable living, with this post as the first in the series. The tone should be informative and inspiring, with a call to follow the series."

12. Create Social Media Posts to Drive Traffic to the Blog

Once the blog post is complete, create social media posts to promote it and drive traffic to your website. Aim for variety in tone, style, or approach to engage different audience segments.

Sample Prompt: "Write three social media posts to promote this blog post on eco-friendly cleaning products. Each post should have a different tone or focus, such as a question to engage followers, a highlight of key benefits, and a call to action to read more on our blog."

Review each social media post to ensure it aligns with the brand's tone and objectives, and save them in your Resource Center for future use.

13. Save and Organize Your Best Outputs in Your AI Resource Center

Save your completed blog post, outline, prompts, and any variations under Writing & Output Examples. This will serve as a template for future blog posts and help build a consistent style.

Common Mistakes to Avoid

1. Not Starting with a Clear Outline

Mistake: Jumping into writing without an outline, leading to a disorganized or unfocused blog post.

Better: Begin with a clear outline to structure the post, including the introduction, key points, and conclusion. An outline provides a roadmap that keeps the content cohesive and aligned with the main topic.

2. Weak Headline or Intro Hook

Mistake: Using a generic headline or introduction that fails to grab attention.

Better: Spend time crafting a compelling headline and a strong introductory hook that draw readers in. A powerful opening sets the tone and encourages readers to continue.

3. Overlooking SEO Keywords

Mistake: Neglecting SEO keywords, which can limit the blog post's visibility in search engines.

Better: Research and incorporate relevant SEO keywords naturally throughout the post. Focus on the headline, introduction, and subheadings to enhance SEO without compromising readability.

4. Overloading with Dense Paragraphs

Mistake: Writing long, dense paragraphs that make the post hard to scan and potentially lose reader interest.

Better: Break up content into short paragraphs, use subheadings, and add bullet points or lists to make the post easy to read. This improves engagement and makes it easier for readers to navigate the content.

5. Forgetting to Include a Call to Action

Mistake: Leaving out a CTA, missing an opportunity to guide readers to the next step.

Better: Conclude with a relevant CTA, such as inviting readers to comment, explore related posts, or check out a product or service. A clear CTA helps drive engagement and keeps readers connected to your brand.

6. Not Reviewing or Editing the AI's Output

Mistake: Using the AI's first draft as-is, which may contain errors or inconsistencies in tone.

Better: Review and edit the AI-generated content to ensure it aligns with your brand voice and meets quality standards. Refine the language, add examples, or personalize the content as needed to make it uniquely yours.

Review the blog post you drafted today. Does it follow a clear structure, starting with an engaging headline and intro? Are SEO keywords integrated naturally, and is the post easy to scan? Check that there's a strong CTA at the end. Refine any areas that could enhance readability, engagement, or alignment with your brand voice.

Reflection for Day 5

What was the purpose of your blog post today, and how well did the AI help you achieve it?

How did experimenting with different hooks or tones impact the post's appeal?

Which keywords did you decide to focus on for SEO, and why?

How comfortable do you feel using the Who/What/How framework to guide the AI in creating blog content?

What other types of blog content would you like to experiment with using AI?

DAY 6

Writing a Frequently Asked Questions (FAQ) Document

Goal:

Use the **Who/What/How framework** to create an FAQ section for your website or marketing materials. Today's activities will help you generate clear, concise answers to common questions and refine the AI's output to match your brand's tone. By the end of today, you'll have a polished FAQ document that addresses key customer concerns and aligns with your brand voice.

Activities for Day 6

1. Start a New Thread and Name It

Open a new thread in ChatGPT and name it **"FAQ Writing Practice.** Starting a new thread helps keep prompts and outputs organized for easy reference.

2. Brainstorm Common Customer Questions

Think about common questions your customers may have about your products, services, shipping, or returns. You can upload past inquiries or notes on recurring questions if you have them to guide the AI's response.

Sample Prompt: "You are a customer support expert **[who the AI is]** helping a company that sells eco-friendly home products **[who the audience is]**. List ten frequently asked questions customers might have about our products, shipping, and return policies **[what the task is]**."

If you already know your most common questions, use them instead and skip to the next step.

3. Write Concise Answers to the FAQs

Using the questions you've identified, prompt the AI to write clear, concise answers. This is where you'll establish a consistent tone—whether friendly, professional, or informative.

Sample Prompt: "Answer these questions in a friendly tone. If you don't know the answer, ask me for clarification."

Once you receive an answer, review it for clarity and tone. Prompt the AI to adjust if it doesn't match your brand's voice.

4. Refine Answers for Specific Concerns

Some FAQs may require additional detail, especially those about product care, return policies, or delivery timeframes. Use the AI to make the responses more detailed or to address specific concerns that customers often have.

Sample Prompt: "Rewrite the answer to 'What is your return policy?' to include information about refunds and return shipping costs."

Once you've reviewed the answer, refine it to ensure it addresses typical follow-up questions.

5. Add Supporting Links and Resources

If certain FAQs would benefit from linking to a related page (e.g., a product care guide or shipping policy), prompt the AI to suggest wording that includes these links naturally.

Sample Prompt: "Add a sentence to the answer for 'How do I care for my product?' that links to a detailed product care guide at [your URL]."

Make sure the links direct customers to the correct resources, and save any answers with helpful links for future use.

6. Incorporate Keywords for SEO

If your FAQ will be on your website, integrating SEO keywords can improve its searchability. Ask the AI to suggest keywords for specific answers and adjust the text to include them naturally.

Sample Prompt: "Suggest SEO keywords for an FAQ section about eco-friendly home products."

Once you've chosen keywords, prompt the AI to add them into the answers where relevant without changing the clarity or tone.

7. Experiment with Tone Options

If you want your FAQ section to feel more approachable, ask the AI to adjust the tone of specific answers to make them warmer or more conversational.

Sample Prompt: "Rewrite the answer to 'What materials are used in your products?' with a more casual and welcoming tone."

Experiment with tone variations until the FAQ answers feel engaging and align with your brand's voice.

8. Create Social Media Posts to Highlight Key FAQs

Promoting FAQs on social media can address common questions publicly and encourage engagement. Prompt the AI to create a few posts that feature important FAQs to share on social platforms.

Sample Prompt: "Create three social media posts highlighting FAQs about our eco-friendly products. Each post should be informative and encourage readers to check out our full FAQ section."

Save these posts in your Resource Center for future sharing on social media to proactively address questions.

9. Ask the AI for Ideas to Use Your FAQs Across Channels

Now that you have your FAQs, explore ways to make it more useful across different platforms and customer interactions. Prompt the AI to suggest creative ways to use FAQs beyond your website.

Sample Prompt: "Suggest ways to use this FAQ section across various channels."

Review the AI's suggestions and note which suggestions make sense. Consider how you might incorporate the FAQ section into email campaigns, social media posts, product packaging, or even printed materials.

10. Save and Organize Your FAQ Section in Your AI Resource Center

Save your FAQs under **Writing & Output Examples**. You can refer back to this section and update it as new questions arise or product details change.

Common Mistakes to Avoid

1. Skipping Customer Research for Common Questions

Mistake: Relying solely on assumptions rather than researching actual customer questions, which can lead to irrelevant FAQs.

Better: Gather real questions from customer service records, social media, or surveys to ensure the FAQ addresses the most pressing issues.

2. Writing Long-Winded Answers

Mistake: Providing overly detailed answers that overwhelm readers, making it harder to find key information.

Better: Keep responses concise, straightforward, and to the point. Use short paragraphs and bullet points where possible to improve readability.

3. Ignoring Tone and Accessibility

Mistake: Writing answers that are overly formal or technical, which may confuse or disengage customers.

Better: Use a friendly, accessible tone that reflects your brand voice and makes information easy to understand, even for those unfamiliar with the topic.

4. Overlooking SEO Opportunities

Mistake: Failing to incorporate SEO keywords, which limits the FAQ's discoverability on search engines.

Better: Integrate relevant keywords naturally into questions and answers to improve search visibility and help more customers find your FAQ section.

5. Missing Links to Additional Resources

Mistake: Not including helpful links to related guides, product pages, or support resources, leaving customers without clear next steps.

Better: Add links to relevant resources, such as product care guides, return policies, or troubleshooting pages, so readers can easily access more detailed information if needed.

6. Neglecting Regular Updates

Mistake: Treating the FAQ as a one-time project, leading to outdated or irrelevant content over time.

Better: Regularly review and update the FAQ to reflect new products, policies, or customer questions. A current FAQ shows customers that you're responsive and attentive to their needs.

Review your FAQ responses. Are they concise, reader-friendly, and written in an accessible tone? Check for any areas where you could add relevant links or SEO keywords to improve searchability and usability. Refine as needed to ensure the FAQ is up-to-date and fully answers common customer questions.

Reflection for Day 6

What was the main focus of your FAQ section, and how well did the AI help you address it?

How did adjusting the tone impact the clarity or friendliness of the answers?

Which FAQs do you think will be most valuable for your customers, and why?

What other types of customer service content would you like to experiment with using AI?

DAY 7

Writing Event Materials

> **Goal:**
> Use the **Who/What/How framework** to create a compelling event announcement. Today's activities will guide you through drafting an announcement and other materials that build anticipation and provide key details to encourage attendance. By the end of today, you'll have a clear, engaging event announcement ready for distribution across multiple channels.

Activities for Day 7

1. Start a New Thread and Name It

Open a new thread in ChatGPT and name it **"Event Announcement Practice."** Starting a new thread helps keep prompts and outputs organized for easy reference.

2. Define the Purpose and Details of the Event

Consider the goal of your event and the key information that attendees need to know, such as the event's purpose, date, location, and any special highlights. Write down these details to help structure your announcement.

Write down the event's purpose and key details:

____

____

____

____

3. **Create a Prompt Using the Who/What/How framework**

Using the **Who/What/How framework**, prompt the AI to create an event announcement that captures excitement and includes essential details. Be specific about who the AI is, who the audience is, what the task is, and how the content should be delivered.

Sample Prompt: "You are an event marketer **[who the AI is]** creating an announcement for eco-conscious customers who are interested in sustainable living **[who the audience is]**. Write an announcement for our upcoming 'Sustainable Home Living Online Workshop' **[what the task is]**. The tone should be upbeat and inviting, and include details about the date, time, location, and RSVP link **[how to deliver the content]**."

4. **Emphasize the Unique Value or Highlights of the Event**

Prompt the AI to focus on what makes this event special, such as exclusive content, guest speakers, or limited-time offers. Highlighting these aspects can increase interest and attendance.

Sample Prompt: "Rewrite the announcement to emphasize the benefits of attending, such as learning eco-friendly home practices, meeting other sustainability-minded individuals, and getting exclusive access to new products."

5. **Include a Clear Call to Action**

Make sure the announcement includes a strong call to action, such as inviting readers to RSVP, register, or sign up for updates.

Sample Prompt: "Add a call to action inviting readers to register for the event and include a link to our RSVP page."

Check that the AI's CTA is clear and compelling. You can prompt it to experiment with different phrasing like "Reserve Your Spot Today!" or "Join Us for This Exclusive Event!"

6. Create Social Media Posts to Promote the Event

Social media is an effective platform for event promotion. Prompt the AI to generate a few posts that highlight different aspects of the event, such as the benefits of attending, a count-down to the event, or a reminder to register.

Sample Prompt: "Create three social media posts to promote the 'Sustainable Home Living Workshop.' Each post should highlight a different aspect, such as the workshop's topic, a guest speaker, and the exclusive discounts for attendees. Include a CTA to sign up."

7. Draft Follow-Up Reminders for Registrants

For attendees who register, follow-up reminders can keep the event on their radar. Ask the AI to draft reminder emails or messages that include the event date, time, and a reminder to save the date.

Sample Prompt: "Write a follow-up reminder email for people who have registered for the Sustainable Home Living Workshop. The tone should be friendly, with a reminder to save the date and details on what to expect."

8. Get Suggestions on Where and How to Promote the Event

Effective event promotion often includes reaching out to relevant channels beyond your own social media or email list. Prompt the AI to suggest places to promote the event and ways to approach each outlet.

Sample Prompt: "Suggest ten places to promote the 'Sustainable Home Living Workshop' and provide tips on how to approach each one. Include options like local eco-friendly blogs, community centers, and online groups focused on sustainability."

Review the AI's suggestions and consider how each channel could be approached. If needed, ask the AI to draft a pitch email or social media message for each recommendation.

9. Experiment with Tone Variations

If your event announcement needs to appeal to different segments, ask the AI to generate variations in tone. For example, you might want a more formal announcement for LinkedIn and a more casual, engaging tone for Instagram.

Sample Prompt: "Rewrite the event announcement in a professional tone for a LinkedIn post."

Experiment with a few tones and save the versions that best fit each platform or audience.

10. Save and Organize Your Event Announcements in Your AI Resource Center

Save your event announcement, social media posts, and reminder messages under Prompts & Templates. This will serve as a reusable template for future events.

Common Mistakes to Avoid

1. Missing Critical Event Details

Mistake: Leaving out essential event information like date, time, location, or registration links, causing confusion and potential drop-offs in attendance.

Better: Double-check that all important details are clearly listed. Consider using a checklist to ensure nothing is missed in your event announcement.

2. Not Emphasizing the Unique Value of the Event

Mistake: Focusing solely on logistic details without explaining why the event is worth attending.

Better: Highlight key benefits, such as networking opportunities, guest speakers, or exclusive content, to attract interest and excitement.

3. Using the Same Messaging Across All Platforms

Mistake: Copying and pasting the same announcement text across different platforms, which can feel impersonal or mismatched.

Better: Customize the tone and structure for each platform to make the message resonate. For example, keep LinkedIn announcements professional, while making Instagram posts more visual and engaging.

4. Forgetting a Clear Call to Action

Mistake: Leaving out a CTA, which can lead to low engagement or unclear next steps for potential attendees.

Better: Add a specific CTA, like "Register Now" or "Save Your Spot," along with a direct link to the registration page. This prompts readers to take action immediately.

5. Not Scheduling Follow-Up Reminders

Mistake: Only sending out the initial announcement, which may not be enough to drive registrations or attendance.

Better: Create a reminder schedule that includes follow-up messages closer to the event date. This helps keep the event on your audience's radar and encourages last-minute sign-ups.

6. Overlooking Opportunities for Cross-Promotion

Mistake: Relying on one channel or audience to drive attendance, potentially missing wider reach.

Better: Cross-promote the event on all relevant platforms and consider reaching out to partners or influencers who can help spread the word.

Review your event announcement materials. Do they clearly include all essential details and a compelling CTA? Are they customized for each platform and audience segment? Set up a schedule for follow-up reminders, and think about any additional channels or partners who could help promote the event. Refine as needed to maximize clarity and reach.

Reflection for Day 7

What was the main goal of your event announcement, and how well did the AI help you achieve it?

How did emphasizing event highlights or benefits impact the announcement's appeal?

Which CTA felt most effective for your audience, and why?

How comfortable do you feel using the Who/What/How framework to guide the AI in creating event announcements?

What other types of event content would you like to experiment with using AI?

WEEK 1 RECAP

BUILDING YOUR AI FOUNDATION

Congratulations on completing your first week working with AI! You've accomplished a significant milestone in your journey toward AI proficiency. Let's review what you've learned and reflect on your progress.

Key Skills and Accomplishments

AI Tool Mastery

- You set up your ChatGPT account and learned to navigate its interface confidently
- Created your AI Resource Center to organize prompts, templates, and outputs
- Mastered thread management and learned to save important conversations for future reference

Prompting Fundamentals

- Learned and practiced the Who/What/How framework for effective prompting
- Discovered how to structure requests for clearer, more accurate responses
- Developed skills in refining and adjusting prompts to improve outputs
- Experienced firsthand how small changes in prompting can lead to better results

Content Creation Basics

- Created your first AI-generated social media posts
- Drafted professional emails with appropriate tone and messaging
- Built basic blog posts and FAQ sections
- Learned to review and refine AI's output for quality and accuracy

Reflection on Progress

Think about where you started versus where you are now:

- How has your comfort level with AI changed?
- Which prompting techniques worked best for you?
- What surprised you about working with AI?
- Which types of content were easiest or most challenging to create?

Common Challenges and Solutions

During this first week, you may have experienced initial uncertainty about how to phrase requests to AI. This is completely normal—learning to communicate with AI is like learning any new skill. You might have received outputs that weren't quite what you wanted, or felt unsure about when to accept AI's responses versus when to ask for refinements. These challenges are natural steps in the learning process, and working through them helps build your confidence and expertise.

Looking Ahead

As we move into Week 2, you'll build on these foundational skills to create more sophisticated content and develop advanced prompting techniques. The basics you've mastered this week—clear communication, organized resource management, and content refinement—will serve as building blocks for more complex AI applications.

Take a moment to celebrate your progress. You've gone from AI novice to someone who can confidently generate basic content using AI tools. That's a significant achievement and sets you up for success in the weeks ahead!

> Now that Week 1 is complete, I will no longer be diagramming the Who/What/How framework in the sample prompts. Make sure you still use that framework with your prompts!

WEEK 2 OVERVIEW

BUILDING BRAND CONSISTENCY AND EFFECTIVE CONTENT CREATION

Welcome to Week 2! Now that you've mastered the basics of AI interaction, we'll focus on creating content that consistently reflects your brand voice while efficiently managing your workflow. This week emphasizes practical applications—you'll develop systems and templates that will streamline your content creation process and ensure everything you produce maintains your brand's unique identity.

What You'll Learn This Week

1. **Brand Voice Development**
   - Creating a comprehensive Brand Persona document
   - Establishing consistent tone and style across platforms
   - Building templates that capture your brand voice
   - Training AI to understand and replicate your style
2. **Content Strategy Enhancement**
   - Developing content pillars that align with your brand
   - Creating targeted messaging for different audience segments
   - Building reusable content frameworks
   - Designing effective CTAs

3. **Workflow Optimization**
   - Setting up Custom GPTs for efficiency
   - Creating standardized prompting templates
   - Developing quality control checklists

Building Your Systems

This week is about moving from individual tasks to systematic approaches. You'll create frameworks and templates that make future content creation faster and more consistent. Think of it as building a content creation machine—once the foundation is in place, you can produce high-quality outputs more efficiently.

The skills you develop this week will help you scale your AI usage effectively. Focus on creating systems that can grow with your needs and adapt to different content types. Remember, the time invested in setting up these frameworks will pay dividends in future content creation efficiency.

DAY 8

Creating a Consistent Brand Voice & Developing Your Brand Persona

Goal:

Learn how to guide the AI to produce content that consistently reflects your brand's unique voice, and create a Brand Persona document. This persona will act as a quick-reference guide that you can copy into any new thread to provide context. By the end of today, you'll have a structured approach to establishing a consistent brand voice and a Brand Persona that serves as your brand's profile.

Activities for Day 8

1. Start a New Thread and Name It

Open a new thread in ChatGPT and name it **"Brand Voice Practice."** This will help keep prompts and outputs organized as you experiment with establishing a consistent brand voice.

2. Define Key Elements of Your Brand Voice

Before creating prompts, define the elements that make up your brand's unique voice and tone. Think about adjectives or phrases that describe how you want your brand to sound, such as "professional but friendly," "casual and fun," or "informative and authoritative." Decide if emojis are part of your voice and, if so, how often and where they should appear (e.g., only in social media posts or selectively in email subject lines).

Write down the elements of your brand voice:

____

____

____

____

3. Create a "For AI Training Purposes" Document in Your AI Resource Center

Build a training document to help the AI mimic your brand voice. This document should include examples of your best-performing content across different formats to ensure the AI has a well-rounded understanding of your style.

If you have your own assets, include your best-performing content pieces across formats, such as:

- **2–3 blog posts**
- **5–10 social media posts**
- **10–15 marketing emails**
- **1 white paper or case study (if available)**

If you don't have these examples or want to mimic another style, gather content examples that fit your desired brand voice and add them to the document. Copy/paste 2–3 samples from blogs, social media posts, or emails that represent the voice you'd like to achieve.

Add the title **"For AI Training Purposes"** at the top of the document, then organize it by content type so the AI can reference the various formats when prompted. Save this document in your AI Resource Center under Brand Voice & Tone Guidelines for ongoing use.

4. Create a Brand Style Guide Document

Alongside the training document, create a Brand Style Guide to help the AI understand specific word choices, spellings, and phrases unique to your brand. This guide will provide instructions on commonly used phrases, words to avoid, custom spellings, and industry-specific acronyms.

Your Style Guide Might Include:

- **Words and Phrases to Avoid:** List overused or generic words and phrases that don't align with your brand's tone.
- **Common Acronyms:** Specify acronyms that are frequently used and recognized by your audience.
- **Special Spellings:** Include custom spellings, hyphenations, or industry-specific terms.

Add "Brand Style Guide" as the title of this document and save it in your AI Resource Center under Brand Voice & Tone Guidelines.

5. Create Your Brand Persona Document and Save it in the AI Resource Center

The **Brand Persona** document provides a clear picture of your company's identity, target audience, and core offerings. This document will help the AI understand your brand's context and will serve as a reference in any new thread.

To Create Your Brand Persona:

1. **Start a New Document** and title it **"Brand Persona."**
2. **Include the following information**:
    - **Company Name and Industry:** State your brand name and primary industry focus.
    - **What You Do:** Briefly describe your main products or services.
    - **Target Audience and Customer Segments:** Define who you serve and describe your main customer segments, including any pain points, needs, or goals they have.
    - **Key Benefits:** Describe the core benefits customers gain from using your products or services.
    - **Unique Selling Points:** Highlight what makes your brand unique in the market.
    - **Product or Service Pricing:** List general price ranges if relevant, especially if they influence purchase decisions.
    - **Additional Context:** Include any other details that would help the AI understand your brand, such as special terms, brand history, or founder values.

Save this document in your AI Resource Center under Brand Persona.

6. Prompting the AI to Review and Clarify

To ensure the AI understands your Brand Persona, copy/paste the document into a prompt and ask it to review for any clarifying questions. This allows you to add any extra details that could help the AI respond accurately in the future.

Sample Prompt:

"You are a brand strategist reviewing our Brand Persona document. Summarize the main details of our brand and ask any clarifying questions you have about our company's mission, audience, or products."

Review the summary and answer any questions the AI generates. Save any adjustments back into your **Brand Persona** document as needed.

7. Maintain Consistency Across Threads or Create Custom Personas

At the time of publication, AI models don't generally retain memory from thread to thread. So every time you start a new thread, you need to remind the model who it is, who you are, and what you're doing. This can get tedious, especially if you switch threads often or have multiple team members using the model. Fortunately, there's a solution if you have access to a paid version of the platform.

To keep your AI aligned with your brand across tasks, you have two options:

1. Upload your Training Document, Brand Voice & Tone Guidelines, and Brand Persona at the start of each new thread.

Or

2. Use a Custom GPT (or equivalent) to retain brand knowledge and streamline your prompts across threads that use that particular model.

Most generative AI platforms offer custom features to create "trained personas" that remember your brand's tone, audience, and background across interactions. As long as you're using the same model across threads, it will remember the context you've given it.

These custom tools have different names across platforms:

- ChatGPT: Custom GPTs
- Claude: Projects
- Gemini: Gems

Using these custom tools, you can create models that retain your brand's guidelines, tone, and purpose across tasks. If the steps below don't match exactly, consult tutorials on your AI platform's help section.

How to Set Up a Custom GPT (or Equivalent) Using the Who/What/How Framework

This setup guide uses the Who/What/How framework to ensure the AI understands its role, tasks, and delivery style. Follow these steps to create a model customized to your brand's needs.

Note: You will need to have a paid version of ChatGPT to create custom models.

Step 1: Access the Custom Builder

Open your AI platform and locate its custom feature. In ChatGPT, the feature is on the left sidebar and labeled "Explore GPTs." Then click the "Create" button at the top right. (Platform design changes constantly, your interface may look slightly different.)

In ChatGPT, you have the option of using the guided builder or working directly through the backend. Let the builder wizard guide you through the process. Just answer the prompts it gives you.

Step 2: In the Builder, Define Who the AI Is and Who the Audience Is

Who the AI Is: Describe the AI's expertise and role(s). For example, you might specify that the AI is a "channel marketer," "cybersecurity content strategist," or "partner enablement specialist with 20 years of experience."

Who the Audience Is: Identify the target audience for this custom model, such as channel partners, IT service providers, or small and medium sized businesses (SMBs).

Example of Who/Who Instructions:

"You are a cybersecurity content strategist and partner enablement expert. You specialize in creating content that helps channel partners understand and sell cybersecurity solutions to SMB clients."

Step 3: Define What the AI's Main Jobs Are

What the AI Does: Specify the primary tasks the AI should focus on, such as writing social media posts, drafting partner emails, or creating product documentation. You can include whatever skills you want. Create your dream assistant! And don't worry, you can always edit these instructions later.

Example of What Instructions:

"Your primary tasks include creating partner-facing email campaigns, drafting event announcements, and writing blog posts on cybersecurity best practices for SMBs."

Step 4: Define How the AI Should Deliver Each Task

How to Deliver: Describe the tone, voice, and style. Provide guidance on vocabulary, formality, and brand-specific preferences.

Example of How Instructions:

"Deliver content in a professional, supportive tone. Use simple, engaging language suitable for channel partners. Avoid jargon and focus on clarity and actionable advice."

Step 5: Upload Training Documents for Background Knowledge

Most custom models allow you to upload relevant documents. At a minimum, upload your Training Document, Brand Voice & Tone Guidelines, and Brand Persona. Add any other resources, such as successful email campaigns or partner program guides.

Step 6: Use the Custom Model for All Relevant Tasks

Use the custom model exclusively for tasks related to the role it was created for. For example, if you create a model for content creation, use it for all email and blog writing. Use the general model for unrelated tasks.

It's best to create several custom models with their own instructions, rather than trying to make one model do lots of different things. You might have a copywriter, data analyst, and campaign strategist, each one highly trained to do their jobs at a high level.

Step 7: Test the Custom GPT with Sample Prompts

Test your setup by prompting the custom model with tasks it was trained for. Since the model already knows and remembers the Who and How context, you only need to specify What you want it to do.

Sample Prompts:

"Write a LinkedIn post announcing our new partner incentive program."

"Create a blog post on how SMBs can strengthen their cybersecurity defenses."

Step 8: Refine Instructions as Needed

Adjust instructions or add specifics if the model's responses need improvement. Most platforms allow you to edit instructions to align better with your needs.

Step 9: Periodic Updates to Keep the Model Aligned

Regularly update the custom model's training documents and instructions as your brand evolves. This ensures the outputs remain consistent with your current brand voice and goals.

Using Your Custom Model with This Book:

Throughout this book, we assume you are NOT using a Custom GPT and guide you through prompts as if you are entering instructions each time. However, if you do create a Custom GPT, you may find prompts easier since your model will already understand your brand's tone, audience, and objectives.

From now on start each new thread by copying your Training Document, Brand Voice Guidelines, and Brand Persona from your AI Resource Center into the prompt box. This will provide the AI with context, ensuring your brand's tone, style, and audience are accurately represented throughout today's activities.

Throughout the rest of this book we'll assume you do NOT have a Custom GPT in place and will guide you through prompts as if you're entering the context each time. However, if you do create a Custom GPT, you'll find the prompts easier to use since that model will understand context across threads.

Common Mistakes to Avoid

1. Skipping the Brand Persona Document

Mistake: Relying on general brand guidelines instead of creating a detailed Brand Persona, which can result in inconsistent messaging.

Better: Develop a clear Brand Persona that includes details on audience segments, brand voice, values, and key messaging. This document ensures everyone working with your brand has a consistent reference.

2. Inconsistent Brand Tone Across Channels

Mistake: Using varying tones on different platforms, leading to a fragmented brand image.

Better: Define and document tone variations for each platform while keeping the overall brand voice consistent. Specify how to adapt tone for LinkedIn, Instagram, Facebook, etc., to suit each audience while maintaining brand unity.

3. Ignoring Audience-Specific Language

Mistake: Using the same messaging for all audience segments, which may miss key nuances that resonate with different groups.

Better: Include tailored language in your Brand Persona for each audience segment, such as younger eco-conscious consumers vs. corporate clients interested in sustainability.

4. Not Documenting Brand-Specific Phrasing or Style Choices

Mistake: Overlooking unique words, phrases, or stylistic elements that reflect your brand, leading to inconsistencies in content.

Better: Create a style guide that includes preferred terminology, brand-specific phrases, and any words or styles to avoid. This keeps language consistent across all content.

5. Skipping Examples of Brand Voice

Mistake: Leaving the AI or your team to interpret tone without specific examples, which can lead to off-brand messaging.

Better: Provide examples of past content that exemplifies your brand voice for each type of content (e.g., blog post, social media update, email). These examples serve as clear references.

6. Failing to Update the Brand Persona Regularly

Mistake: Treating the Brand Persona as a static document, which can lead to outdated messaging as the brand evolves.

Better: Schedule regular updates to the Brand Persona to reflect any shifts in audience preferences, tone adjustments, or new brand guidelines.

Review your Brand Persona and Brand Style Guide documents. Do they include specific language, examples, and guidelines for each audience and platform? Is the tone consistent across different channels? If any sections feel unclear, refine them or add examples to ensure they provide comprehensive guidance for anyone creating content for your brand.

Reflection for Day 8

What elements of your brand voice felt the most unique, and how well did the AI capture them?

How comfortable do you feel using the Who/What/How framework to prompt the AI?

What details did you add to your Brand Persona that you think will help the AI better understand your brand?

How will the Brand Persona and other documents help streamline future content creation?

DAY 9

Audience Segmentation and Personalization

Goal:

Learn how to create personalized content for different audience segments using the **Who/What/How framework**. Today's activities will help you identify key audience segments, craft messages that speak directly to each group, and refine prompts to generate content that feels personal and relevant. By the end of today, you'll be able to guide the AI to create content that better connects with each unique segment of your audience.

Activities for Day 9

1. Start a New Thread and Name It

Open a new thread in ChatGPT and name it **"Audience Personalization Practice."** This will keep your prompts and outputs organized as you work on segmenting and personalizing content for different audience groups.

Reminder: If you haven't set up a Custom GPT yet, start each new thread by copying your **Training Document, Brand Voice Guidelines, and Brand Persona** from your AI Resource Center into the prompt box. This will provide the AI with context, ensuring your brand's tone, style, and audience are accurately represented throughout today's activities.

2. Define Your Key Audience Segments

Consider the different groups you typically address in your marketing. For today's practice, we'll focus on two distinct audience segments:

- **Eco-conscious young adults** who are interested in sustainable lifestyle choices
- **Corporations seeking to go green** as part of their environmental or corporate social responsibility (CSR) initiatives

Write down any additional audience segments you'd like to address in the future:

____

____

____

____

____

____

____

____

3. Create a Prompt to Identify Segment-Specific Needs and Preferences

Using the **Who/What/How framework**, prompt the AI to create a brief profile of each segment, including their interests, needs, and preferences. This will help ensure that your messaging is relevant and speaks to each group's unique motivations.

Sample Prompt: "You are a market research analyst identifying customer preferences for eco-friendly products. Briefly describe the main interests and priorities of two key segments: eco-conscious young adults and corporations seeking to go green. Summarize in bullet points."

Use the profiles generated to guide your content, ensuring each segment's needs and values are top of mind.

4. Practice Writing Tailored Content for Each Segment

Prompt the AI to create content specifically for each segment based on the profile. This could be a social media post, an email, or a product description with language and points that resonate with that group.

Sample Prompt: "Write a social media post promoting our reusable kitchenware for eco-conscious young adults. Make it engaging and emphasize the environmental impact of small changes."

Experiment with different tones, phrases, or CTAs based on each segment's preferences. Repeat this process for each segment to practice consistent personalization.

5. Create Targeted Email Subject Lines

Tailor subject lines for emails based on the audience segment to improve open rates and engagement. Prompt the AI to generate subject lines that appeal to the specific interests or motivations of each group.

Sample Prompt: "Suggest three email subject lines promoting eco-friendly office supplies to corporations seeking to improve their sustainability practices."

Review the options, choosing the subject lines that best align with each segment's values.

6. Customize Calls-to-Action for Each Segment

Different audiences may respond better to certain types of CTAs. Ask the AI to suggest CTAs that would specifically appeal to each audience segment.

Sample Prompt: "Create a CTA for corporations to explore our sustainable office product line. Make it professional and action-oriented."

If you find certain phrases or CTAs are more appealing for specific segments, note them for future use in the AI Resource Center under Prompts & Templates.

7. Translate Content for a Multilingual Audience

If your business is expanding to reach multilingual customers, practice translating content for consistency across languages. Start by prompting the AI to translate a piece of content into Spanish and French, and review the output for clarity and tone.

Sample Prompt: "Translate the following product description into Spanish and French: 'Our eco-friendly water bottles are made with sustainable materials and are perfect for on-the-go hydration. Make a positive impact with every sip!'"

If you have team members or contacts who speak the target languages, have them review the translations for accuracy.

8. Adjust Tone and Language for Segment Preferences

Language preferences, tone, and even terminology can vary across segments. Practice asking the AI to modify tone and vocabulary to match what resonates with each audience.

Sample Prompt: "Rewrite this product description in a more formal, informative tone for corporations interested in sustainability initiatives."

Experiment with adjusting tone and language for other segments to see which approaches best fit each group.

9. Save Custom Prompts for Future Use

Save your favorite prompts and audience-specific messaging examples in the **AI Resource Center**. Having these tailored prompts and examples on hand will streamline future content creation for each segment.

Copy and paste your effective prompts, customized CTAs, and tailored language samples for each segment into your Resource Center.

Common Mistakes to Avoid

1. Not Defining Audience Segments Clearly

Mistake: Overlooking the unique characteristics and preferences of different audience segments, which can lead to generalized messaging.

Better: Clearly outline key audience segments, such as eco-conscious young adults or corporate clients interested in sustainability. Define each segment's needs, interests, and preferred messaging style to guide more personalized content creation.

2. Using the Same Message Across All Segments

Mistake: Sending identical content to all segments, which risks not resonating with any specific group.

Better: Tailor your messaging to each segment. Create unique value propositions or angles that speak directly to the interests of each group, making the content feel personal and relevant.

3. Overlooking Segment-Specific Calls-to-Action

Mistake: Using a generic CTA for all audiences, which may not motivate specific groups to act.

Better: Customize CTAs to reflect each segment's needs or goals. For instance, young adults may respond well to "Explore Sustainable Choices," while corporate clients might prefer "Discover Solutions for Sustainable Business."

4. Ignoring Tone Variations for Different Segments

Mistake: Applying a one-size-fits-all tone across diverse audience groups, which can make the message feel mismatched or impersonal.

Better: Adjust the tone to fit each segment's preferences. Young customers may respond well to a conversational, friendly tone, while a more formal and informative tone may better suit corporate audiences.

5. Not Testing or Reviewing Segment-Specific Content

Mistake: Skipping the testing phase, which can result in missed insights about how each segment engages with content.

Better: Test different versions of content for each audience segment and review performance data regularly. Adjust based on what resonates most with each group to improve future engagement.

6. Neglecting to Document Insights

Mistake: Failing to document effective prompts, language, and segment preferences, making it harder to replicate successful content.

Better: Save successful messaging examples, prompts, and preferences for each segment in your AI Resource Center. This serves as a valuable reference for future content tailored to each audience group.

Review the segment-specific content you created today. Does each segment's content feel personalized and relevant, with a clear CTA that aligns with their interests? Check that the tone is appropriate for each audience and adjust as needed. Refine your documentation with any effective prompts or phrases you found useful for specific segments.

Reflection for Day 9

Which audience segments did you focus on today, and how well did the AI capture each group's unique preferences?

How comfortable do you feel prompting the AI to create content that resonates with specific audience segments?

What adjustments in tone, language, or content were most effective for different segments?

Which personalized content pieces felt most aligned with your brand and segment needs?

What other audience segments or personalization techniques would you like to experiment with using AI?

DAY 10

Repurposing Content

Goal:

Learn how to repurpose existing content into various formats using the **Who/What/How framework**. Today's activities will guide you in transforming blog posts, emails, or other written content into social media posts, email newsletters, infographics, and more. By the end of today, you'll know how to use AI to reformat and optimize content for different channels, maximizing its value and extending its reach.

Activities for Day 10

1. Start a New Thread and Name It

Open a new thread in ChatGPT and name it **"Content Repurposing Practice."** This will keep prompts and outputs organized as you experiment with repurposing content into various formats.

> **Reminder:** *If you haven't set up a Custom GPT yet, start each new thread by copying your* **Training Document, Brand Voice Guidelines, and Brand Persona** *from your AI Resource Center into the prompt box. This will provide the AI with context, ensuring your brand's tone, style, and audience are accurately represented throughout today's activities.*

2. Choose a Piece of Existing Content to Repurpose

Select a high-performing or cornerstone piece of content, such as a blog post, email, or white paper. This will be your base material for today's repurposing activities.

Write down the content type you'll be repurposing and its main topic or title:

____

____

____

____

3. Upload the Content Document and Summarize the Original Content

Upload the content document to help the AI understand the material you're working with. Then prompt it to generate a brief summary, focusing on key points. This summary will help simplify your starting point for each new format and keep the main ideas consistent across repurposed versions.

Sample Prompt: After uploading, type: "Summarize this blog post on sustainable kitchen practices in two to three sentences."

Use the summary as a reference point for creating variations while keeping the main message intact.

4. Create Social Media Posts from the Original Content

Ask the AI to turn the original content into a series of social media posts, each with a unique angle or emphasis. This is a great way to extend the life of a single piece of content across platforms.

Sample Prompt: "Create three social media posts based on this blog post about sustainable kitchen practices. Each post should highlight a different benefit of going green in the kitchen."

Experiment with posts for different social platforms (e.g., LinkedIn, Instagram, X).

5. Generate an Email Newsletter Based on the Content

Turn the content into a short email newsletter to share with your audience. Include a brief intro, key points, and a CTA.

Sample Prompt: "Write an email newsletter promoting the benefits of sustainable kitchen practices. Include a brief introduction, three main points, and a CTA to read the full blog post on our website."

Adjust the tone to be engaging and inviting, then save the draft for future use.

6. Transform the Content into a Visual Format

Visuals can be powerful tools for engaging readers. Prompt the AI to outline an infographic, checklist, or slide deck based on the main points of your content.

Sample Prompt: "Create an outline for an infographic on sustainable kitchen practices, with three sections: Reducing Waste, Choosing Eco-Friendly Materials, and Conserving Water and Energy."

Use the outline to design an infographic or visual aid with any design tools you have available.

7. Combine Two Pieces of Content into One Cohesive Format

If you have two related content pieces, prompt the AI to combine them into one larger format, such as a comprehensive guide or a long-form article.

Sample Prompt: "Combine this blog post on sustainable kitchen practices with our email about eco-friendly kitchen tools to create a single long-form article. Include a summary, key sections, and a concluding CTA."

Review the combined piece and refine it as necessary.

8. Reformat for a Short-Form Blog Post or Article

If the original content is long-form, prompt the AI to turn it into a shorter, more concise version for a quick-read blog post or article.

Sample Prompt: "Condense this blog post on sustainable kitchen practices into a 300-word article, focusing on the top three benefits."

Use this condensed article for platforms where shorter content is better suited, such as email or social media-linked articles.

9. Translate the Repurposed Content into Other Languages

If you're working with multilingual audiences, prompt the AI to translate your repurposed content into relevant languages, such as Spanish and French.

Sample Prompt: "Translate this social media post about sustainable kitchen practices into Spanish and French."

Make sure a native speaker reviews it for accuracy.

10. Transform Content into an Interactive Format

Engage your audience by repurposing content into a quiz, checklist, or other interactive format. This can be especially useful for topics where readers may enjoy testing their knowledge or exploring practical steps.

Sample Prompt: "Create a 5-question quiz based on our blog post about sustainable kitchen practices. Each question should provide an informative answer to educate readers."

Use the quiz or checklist in email newsletters, on your website, or in social media stories to increase engagement.

Common Mistakes to Avoid

1. Choosing Content to Repurpose Without a Strategic Goal

Mistake: Repurposing content simply to generate more posts without a clear objective, which can lead to repetitive or unfocused messaging.

Better: Identify the goal for each repurposed piece, such as increasing brand awareness, educating new audiences, or driving conversions. A focused goal will guide you on how to adapt the content most effectively.

2. Over-Reliance on Direct Copying Instead of Tailoring Content

Mistake: Copying content directly from one platform to another without adapting it to fit the new format, audience, or tone.

Better: Adjust each piece for its specific platform and audience. For example, a blog post repurposed for social media should be more concise, with an eye-catching image or graphic to draw attention.

3. Ignoring Platform Norms and Preferences

Mistake: Not tailoring content to fit the norms of each platform, which can lead to low engagement.

Better: Adjust repurposed content to align with platform norms—use hashtags on Instagram, professional tone on LinkedIn, and shorter text on Twitter/X. Tailoring format and style improves resonance on each platform.

4. Missing Opportunities to Update or Refresh Content

Mistake: Repurposing older content without refreshing or updating information, which can make it feel stale or outdated.

Better: Review each piece to see if it needs a fresh angle or updated statistics. Adding current insights or new visuals can make repurposed content feel relevant and timely.

5. Not Utilizing AI to Generate Format-Specific Variations

Mistake: Manually trying to rephrase content for each platform, which can be time-consuming and lead to inconsistencies.

Better: Use AI to quickly generate variations for different formats. For example, ask AI to condense a blog into a tweet thread or create an Instagram-friendly caption. This can streamline the process and help maintain brand consistency.

6. Forgetting to Test and Optimize Repurposed Content

Mistake: Assuming that repurposed content will work as effectively in the new format as it did in the original.

Better: Monitor engagement on repurposed pieces and use analytics to see which formats resonate most. This feedback helps you refine future repurposing strategies for each platform.

Review the repurposed content you created today. Did you adjust each piece to fit the norms and preferences of its new platform? Use AI to create at least one alternative format that you haven't yet tried, like a condensed LinkedIn summary of a blog post or a visual listicle for Instagram. Track engagement to see what works best and note any insights in your AI Resource Center for future reference.

Reflection for Day 10

Which piece of content did you choose to repurpose, and why?

How effective was the AI in generating repurposed content that matched the original message?

Which formats (e.g., social media posts, email newsletters, infographics) felt most effective for reaching your audience?

How comfortable do you feel using the AI to transform one piece of content into multiple formats?

What other types of content would you like to experiment with repurposing using AI?

DAY 11

Creating Data-Driven Content

Goal:

Learn how to use AI to create content that includes data, statistics, and relevant sources. Today's activities will guide you through prompting the AI to find and incorporate reliable data points, suggest source citations, and provide insights to back up your key messages. By the end of today, you'll be able to create credible, data-rich content that resonates with your audience.

Note:

Data privacy and security are critical. Please do not upload any content that includes personally identifiable information (PII) or sensitive data, unless you are working on a private, secure server. Use only anonymized or hypothetical data when practicing these exercises.

Activities for Day 11

1. Start a New Thread and Name It

Open a new thread in ChatGPT and name it **"Data-Driven Content Practice."** This will keep prompts and outputs organized as you work on incorporating data and statistics into your content.

Reminder: *If you haven't set up a Custom GPT yet, start each new thread by copying your* **Training Document, Brand Voice Guidelines, and Brand Persona** *from your AI Resource Center into the prompt box. This will provide the AI with context, ensuring your brand's tone, style, and audience are accurately represented throughout today's activities.*

2. Identify Key Points That Need Data Support

Select a topic or piece of content you'd like to back up with data. Identify which main points, arguments, or claims would be stronger with supporting data, statistics, or examples.

Upload the document and prompt the AI to identify areas where supporting data could strengthen the content.

Sample Prompt: After uploading, type: "Review this content and identify where supporting data, statistics, or sources could be helpful."

Write down 2–3 key points the AI suggests to support with data:

____

____

____

____

3. Prompt the AI to Suggest Relevant Data or Statistics

Using the **Who/What/How framework**, prompt the AI to suggest data or statistics to support each key point. Be specific about the type of data you're looking for, and ask the AI to provide source citations wherever possible.

Sample Prompt: "Provide recent statistics on consumer interest in eco-friendly kitchen products to support our blog post on sustainable kitchen practices. Include source citations where available."

After receiving the data, review the sources provided and verify accuracy by checking original citations.

4. Analyze Hypothetical Survey Data and Draw Insights

If you don't have your own data, use the following hypothetical survey data to practice. Prompt the AI to interpret the data and draw relevant insights.

- **Survey on Eco-Friendly Product Preferences (Hypothetical Data):**
  - 60% of consumers prefer eco-friendly kitchen products over conventional products.
  - 45% are willing to pay up to 20% more for eco-friendly products.
  - 35% feel that eco-friendly products positively impact their overall well-being.

Sample Prompt: "Analyze this survey data on consumer attitudes toward eco-friendly products and summarize the main trends in three sentences."

Use these insights to enrich your content by adding context or highlighting important patterns.

5. Create a Data Summary for Social Media

Data-driven posts can increase engagement on social media by offering value and credibility. Prompt the AI to create a social media post summarizing the key data points, using visuals or bullet points if appropriate.

Sample Prompt: "Create a LinkedIn post summarizing recent data on the growth of eco-friendly products, including two main statistics and a CTA to read our full report."

Review the summary, and request refinements as needed.

6. Add Data and Citations to Existing Content

Select a piece of content you've already written, such as a blog post or email. **Upload the content document first**, then prompt the AI to incorporate relevant data points and citations to enhance its credibility.

Sample Prompt: After uploading, type: "Add three data points on sustainable kitchen practices, with source citations, to this blog post."

Verify all citations provided and save the updated, data-supported content.

7. Write a Data-Focused Email Newsletter

Create a data-driven email newsletter by prompting the AI to include a few key statistics or insights related to your topic. This approach can increase credibility and reader engagement.

Sample Prompt: "Write an email newsletter introducing our latest report on eco-friendly kitchen products. Include two key statistics and a CTA to download the full report."

If the email aligns with your brand and engages readers, consider saving it as a template in your email marketing software.

8. Create a Chart and Customize It to Match Brand Colors

Visual representations can make data more digestible and appealing. Use the following sample data to practice asking the AI to create a chart, and prompt it to customize the chart with your brand colors for consistency.

Sample Data:

- 2019: 15% growth in eco-friendly product sales
- 2020: 22% growth
- 2021: 35% growth
- 2022: 40% growth

Sample Prompt: "Create a line chart visualizing the annual growth in eco-friendly product sales based on this data, and change the chart colors to match our brand colors: green (#28A745) and light grey (#E2E8F0)."

Save the chart and prompt in your **AI Resource Center** for future reference in visual content.

9. Add a Data-Based Quiz or Poll to Your Content

Interactive content can drive engagement by involving your audience in data-based activities. Prompt the AI to create a short quiz or poll question based on data insights from your topic.

Sample Prompt: "Create a 3-question quiz on recent trends in eco-friendly products, using data to support each answer. Include explanations for each correct answer."

Use the quiz on social media, in emails, or on your website to boost engagement and make data-driven content fun for your audience.

10. Save Data-Driven Content and Prompts

Save your data-rich content, prompts, and any related visuals in the **AI Resource Center**. This will give you ready-to-use examples and a resource for future data-based content projects.

Copy and paste successful data-driven prompts, content samples, and visual outlines into your Resource Center to create a toolkit for data-supported content.

Common Mistakes to Avoid

1. Relying on AI Data Without Verification

Mistake: Assuming all data provided by AI is accurate, which can lead to incorrect or misleading information in content.

Better: Always verify data and sources provided by the AI, especially if it's for public-facing content. Cross-check important data with credible sources to ensure accuracy.

2. Requesting Data Without Providing Clear Parameters

Mistake: Asking AI to add "relevant data" without specifying timeframes, locations, or topics, leading to overly general or off-target information.

Better: Be specific in your prompts about the type of data needed, including details like year, geographic region, or industry. Clear parameters will result in more relevant and useful data.

3. Overloading Content with Too Many Data Points

Mistake: Adding too much data, which can overwhelm the reader and detract from the main message.

Better: Focus on a few high-impact statistics or insights that support your message. Use data selectively to add credibility without overcomplicating the content.

4. Not Considering Visual Representations for Data

Mistake: Presenting all data in text form, which can make it harder for readers to grasp key points.

Better: Ask AI to suggest or create charts, graphs, or infographics to illustrate key data points. Visual representations make data easier to digest and can enhance engagement.

5. Using Generic Phrasing Instead of Data-Driven Insights

Mistake: Leaving out specific insights or trends when presenting data, which can make the content feel flat or generic.

Better: Prompt the AI to interpret data by identifying trends, comparisons, or projections. Adding commentary about what the data means for your audience will make the content more valuable and engaging.

6. Ignoring Data-Relevant SEO Keywords

Mistake: Overlooking SEO opportunities when adding data, which can reduce the content's searchability.

Better: Integrate SEO-friendly keywords related to the data or topic throughout the content. For instance, use terms like "eco-friendly trends 2024" or "small business marketing statistics" to improve discoverability.

Go back through today's data-driven content. Did you verify key data points? Check that the data is presented in a clear, engaging way—whether through visuals, insights, or succinct text. Use AI to create a visual representation of one key data point and refine any phrases to include relevant keywords. Update your AI Resource Center with any data or prompts that worked well.

Reflection for Day 11

What topic did you focus on for creating data-driven content, and how well did the AI support your needs?

How effective was the AI in finding relevant data and providing source citations?

Which data-driven content formats (e.g., social media, infographic, email) resonated most with your brand?

How comfortable do you feel using the AI to incorporate data and statistics into your content?

What other data-supported content ideas would you like to experiment with using AI?

DAY 12

Creating AI-Powered Content Calendars

Goal:
Learn how to use AI to create and manage a content calendar tailored to your brand's goals and audience. Today's activities will guide you in generating content ideas, organizing them into a calendar, and planning ahead for consistent posting. By the end of today, you'll have a structured approach to managing content that keeps your messaging fresh and aligned with your brand strategy.

Activities for Day 12

1. Start a New Thread and Name It

Open a new thread in ChatGPT and name it **"Content Calendar Practice".** This will help you keep all prompts and outputs organized as you build out your content calendar.

> **Reminder:** *If you haven't set up a Custom GPT yet, start each new thread by copying your* **Training Document, Brand Voice Guidelines, and Brand Persona** *from your AI Resource Center into the prompt box. This will provide the AI with context, ensuring your brand's tone, style, and audience are accurately represented throughout today's activities.*

2. Define Your Monthly or Quarterly Content Goals

Consider what you want to achieve with your content over the next month or quarter. Goals might include increasing brand awareness, driving website traffic, or promoting specific products or services.

Write down 2–3 main goals for your content calendar:

____

____

____

____

3. Upload Your Content Pillars or Prompt the AI to Generate New Ones

If you already have established content pillars (key themes that align with your brand's messaging), upload them to guide the AI in generating content ideas. If you don't have content pillars yet, prompt the AI to suggest pillars based on your brand's mission, values, and audience.

Sample Prompt (if you need new pillars): "Suggest content pillars for a brand focused on eco-friendly kitchen products, targeting sustainability-minded consumers."

Review the AI's suggestions and choose the pillars that best reflect your brand's voice and goals.

4. Generate Content Ideas for Each Goal

Using the **Who/What/How framework**, prompt the AI to brainstorm content ideas that align with each goal and your content pillars. These ideas should be varied, covering different formats (e.g., blog posts, social media posts, email newsletters) to provide a well-rounded approach.

Sample Prompt: "You are an expert social media content specialist brainstorming content for young, environmentally-conscious online shoppers. Draft 10 content ideas for a social media campaign focused on increasing awareness about of eco-friendly products. The tone should be engaging and informative."

Review the AI's suggestions, choose the ones that align best with your goals, and organize them by format or theme.

5. Create a Month-Long Content Plan in Table Format

Organize the selected content ideas into a month-long content plan in table format, covering various platforms. Start by designating specific days for each type of content (e.g., social media posts on Mondays, blog posts on Thursdays) and prompt the AI to display the plan in a table for clarity.

Sample Prompt: "Organize these content ideas into a one-month content calendar in table format, with columns for date, content type, platform, and topic."

Adjust the AI's table layout if necessary, and save it for future reference.

6. Plan for Seasonal or Timely Content

Incorporate any seasonal or timely content relevant to your brand, such as holiday promotions, awareness days, or industry-specific events. Prompt the AI to suggest relevant topics based on upcoming dates.

Sample Prompt: "Suggest content ideas for Earth Day to promote our sustainable kitchen products on social media."

Add these seasonal pieces to your calendar, filling in the calendar with timely, relevant content.

7. Generate Supporting Visual Ideas for Social Media

Visuals are crucial for social media posts. Ask the AI to suggest ideas for visuals that could accompany your content, such as infographics, photos, or graphics that emphasize key messages.

Sample Prompt: "Suggest five visual ideas to accompany posts promoting our eco-friendly kitchenware on Instagram."

Organize these visual ideas in your content calendar or link them to specific posts for easy planning.

8. Build Out Content Themes for Each Week

For a cohesive flow, create weekly themes or content clusters that align with your goals and content pillars. These themes could be based on product features, educational insights, customer testimonials, or industry trends.

Sample Prompt: "Develop a weekly theme for each of the next four weeks to focus on different aspects of sustainable living, including energy savings, water conservation, waste reduction, and eco-friendly materials."

Review the AI's weekly themes and add them to the calendar. You can save these themes in your AI Resource Center as inspiration for future planning.

9. Schedule Engagement and Follow-Up Posts

Effective content calendars include engagement touchpoints, like follow-up posts or responses to popular topics. Ask the AI to suggest engagement posts or questions to foster audience interaction and extend the life of each main post.

Sample Prompt: "Suggest engagement posts to follow up on our Earth Day posts, such as questions or polls related to eco-friendly practices."

Add these engagement posts to the calendar on specific days following main posts, ensuring that each topic stays active and engaging.

Common Mistakes to Avoid

1. Creating a Content Calendar Without Defining Clear Goals

Mistake: Jumping into content planning without specific goals, resulting in a calendar that lacks strategic direction.

Better: Set measurable goals for the month or quarter, such as increasing engagement or promoting a specific product. These goals guide the focus of your content and make it easier to measure success.

2. Ignoring Audience Preferences in Content Planning

Mistake: Planning content based solely on brand priorities, without considering what your audience finds valuable or engaging.

Better: Include audience-focused themes in your calendar, based on feedback, popular topics, or analytics. Content that aligns with audience preferences is more likely to drive engagement and reach.

3. Overloading the Calendar with Too Many Themes

Mistake: Trying to cover too many topics in one month, which can dilute the brand message and overwhelm both the team and the audience.

Better: Stick to a few core themes that support your main goals, allowing for more focused, consistent messaging. Limiting themes helps streamline content production and maintains a cohesive brand narrative.

4. Not Optimizing for Seasonal or Timely Content

Mistake: Overlooking seasonal opportunities, which can make content feel out of touch or generic.

Better: Incorporate seasonal themes or events that are relevant to your audience, such as holidays, industry events, or trending topics. Timely content helps keep your brand relevant and can boost engagement.

5. Neglecting to Schedule Content for Different Platforms

Mistake: Planning content without specifying where each piece will be published, which can lead to inconsistent posting across platforms.

Better: Note specific platforms for each post in your content calendar, and tailor the content to fit each platform's tone, format, and audience expectations.

6. Failing to Review and Adjust the Calendar Regularly

Mistake: Treating the calendar as static, which can lead to outdated or misaligned content if goals or audience needs shift.

Better: Schedule regular reviews to adjust the calendar based on analytics and current priorities. Flexibility ensures that content remains relevant and effective.

Try this: Review your content calendar. Does it have clear goals, platform-specific plans, and audience-focused themes? Check for any seasonal opportunities you may have missed, and ensure there's a balance between your brand priorities and what's valuable to your audience. Adjust as needed, and note any prompts or themes that worked well in your AI Resource Center.

Reflection for Day 12

What goals did you focus on for this content calendar, and how well did the AI support your planning needs?

How effective was the AI in generating content ideas that aligned with your brand goals?

Which content types (e.g., social media, blog posts, email) felt most valuable for your audience in this plan?

How comfortable do you feel using the AI to build and organize a comprehensive content calendar?

What other planning elements or tools would you like to experiment with using AI?

DAY 13

Conducting an AI-Powered Competitor Analysis

Goal:

Learn how to use AI to conduct a basic competitor analysis, identifying strengths, gaps, and potential opportunities in your market. Today's activities will guide you in evaluating competitor content, analyzing industry trends, and extracting insights to inspire your own content strategy. By the end of today, you'll have a clearer understanding of your competitive landscape and new ideas for differentiating your brand.

Activities for Day 13

1. Start a New Thread and Name It

Open a new thread in ChatGPT and name it **"Competitor Analysis Practice"**. This will help keep your prompts and outputs organized as you explore competitor insights.

Reminder: If you haven't set up a Custom GPT yet, start each new thread by copying your **Training Document, Brand Voice Guidelines, and Brand Persona** from your AI Resource Center into the prompt box. This will provide the AI with context, ensuring your brand's tone, style, and audience are accurately represented throughout today's activities.

2. Identify 3–5 Key Competitors in Your Industry

List three to five of your main competitors. These can be brands that offer similar products or services, have overlapping audiences, or are recognized leaders in your industry.

List your top competitors:

____

____

____

____

____

3. Copy/Paste Content or Link to Competitor Content for Analysis

If you have samples of your competitors' content, copy and paste them into the AI for analysis. Alternatively, you can include the URL of the competitor's blog or social media profile.

Note: Sometimes, the AI may be restricted from reading a URL directly from certain sites. If that happens, describe the content briefly, or try pasting a text excerpt for analysis.

Sample Prompt: "You are a market analyst evaluating blog posts from Competitor X. Identify three content strengths and three areas where we could improve on their approach. The tone should be objective and focused on actionable insights."

Review the AI's feedback to identify where your brand might add unique value or fill a gap in the industry.

4. Conduct a Content Gap Analysis Using Competitor URLs

A content gap analysis helps identify topics that competitors may be missing or underexploring. To do this, upload URLs for competitor blogs or social media profiles and prompt the AI to evaluate which topics are covered extensively and which are underrepresented.

Sample Prompt: "You are a content strategist reviewing recent blog topics from Competitor X and Competitor Y. Identify any content gaps or topics related to eco-friendly home products that they haven't covered. Please keep the tone professional and objective."

Use the AI's suggestions to create a list of unique topics you could explore for your own content.

5. Analyze Social Media Strategy and Engagement

Understanding how competitors engage with their audience on social media can offer valuable insights. Ask the AI to evaluate your competitors' social media posts and identify the types of content that generate high engagement.

Sample Prompt: "You are a social media strategist analyzing recent social media posts from Competitor X. Identify the content types that generate the most engagement and suggest one way we could improve on their approach. Keep the tone constructive and focused on actionable insights."

Note the strategies that resonate with your audience, and consider how to adapt them with your brand's unique angle.

6. Generate Inspiration from Competitor Content

Competitor analysis can reveal valuable inspiration for content creation. Prompt the AI to suggest content ideas based on high-performing competitor topics, ensuring each idea is adjusted to fit your brand's voice and values.

Sample Prompt: "You are a content creator generating unique content ideas for eco-conscious consumers based on Competitor X's top-performing posts on eco-friendly kitchen tips. Suggest three content ideas that align with our brand's voice. The tone should be fresh, informative, and aligned with our eco-friendly mission."

Choose the most relevant ideas and add them to your content calendar.

7. Define Your Brand's Unique Points of View on Common Topics

Creating a distinct point of view (POV) for common industry topics can differentiate your brand and resonate with your audience. Prompt the AI to help you identify unique angles or stances your brand could take on common topics.

Sample Prompt: "You are a brand strategist helping our eco-friendly brand stand out in a crowded market. Suggest unique perspectives we could take on common eco-friendly home product topics, such as reducing waste, sustainable materials, and water conservation. The tone should be bold and distinctive, with a clear emphasis on originality."

8. Identify Competitor Weaknesses and Areas for Differentiation

Analyzing weaknesses or areas competitors may be overlooking can help you position your brand to stand out. Prompt the AI to highlight any evident weaknesses or areas where competitors may lack depth.

Sample Prompt: "You are a competitive analyst reviewing Competitor Y's blog on sustainable home practices. Identify any weaknesses or missing elements, and suggest ways we could cover this topic in more depth. The tone should be constructive and focused on opportunity."

Use the AI's suggestions to outline potential areas for deeper content, which you can add to your brand's unique positioning.

9. Track Industry Trends and Audience Preferences

Analyzing industry trends is a key part of competitor analysis. Ask the AI to provide insights into current trends within your industry and how competitors are responding to these trends.

Sample Prompt: "You are a market research analyst identifying three current trends in the eco-friendly home products industry. Describe how competitors are addressing each trend and any potential opportunities we could explore. The tone should be professional and focused on strategic insights."

Use these trends to adjust or expand your content strategy to keep it relevant to audience interests.

10. Compile Findings into a Competitor Analysis Report

Organize your insights into a concise report. Prompt the AI to help summarize key points, including competitor strengths, content gaps, engagement strategies, and potential opportunities.

Sample Prompt: "You are a market analyst creating a competitor analysis summary. Summarize key findings on competitor strengths, content gaps, and potential areas for our brand to differentiate. The tone should be clear, concise, and focused on actionable insights."

Save this report for future strategy decisions.

Common Mistakes to Avoid

1. Not Identifying Key Competitors Before Analysis

Mistake: Starting a competitor analysis without selecting specific competitors, leading to a vague or inconsistent approach.

Better: List 3–5 key competitors whose strategies you want to evaluate. Knowing who to focus on helps you extract relevant insights and meaningful comparisons for your brand.

2. Overlooking Gaps in Competitor Content

Mistake: Focusing only on what competitors do well without identifying gaps, which misses opportunities to differentiate your brand.

Better: Look for underrepresented topics or areas competitors aren't covering thoroughly. Filling these content gaps can set your brand apart and meet needs that others overlook.

3. Using General or Vague Prompts for Competitor Analysis

Mistake: Giving AI broad instructions like "analyze competitor content," which can result in unfocused insights.

Better: Provide specific prompts, such as "Identify three strengths in Competitor X's blog strategy" or "What topics are missing from Competitor Y's social media content?" This focused approach yields more actionable insights.

4. Ignoring Social Media Engagement Analysis

Mistake: Only reviewing competitor content without analyzing engagement levels, which can lead to misunderstandings about what actually resonates with audiences.

Better: Check engagement metrics like likes, comments, and shares on competitor posts to see which types of content drive the most interaction. This insight can guide your own content strategy.

5. Copying Competitor Tactics Instead of Adapting Them

Mistake: Trying to replicate competitor strategies exactly, which can make your brand feel unoriginal or out of sync with its audience.

Better: Use competitor insights as inspiration, but adapt tactics to fit your brand's voice, audience, and unique selling points. Differentiation is key to standing out in a competitive market.

6. Not Documenting Competitor Insights in an Organized Way

Mistake: Collecting notes or observations on competitors but failing to save them in an organized, accessible format.

Better: Create a Competitor Insights document, with sections for strengths, weaknesses, content gaps, and potential opportunities. This document becomes a valuable reference for future content and strategy planning.

Review the competitor insights you gathered today. Did you identify specific strengths, content gaps, and engagement trends? Refine your notes to ensure they're actionable and organized. Use AI to generate an outline for a Competitor Insights document if you haven't yet created one, and update it with any new insights that could inform your strategy.

Reflection for Day 13

Which competitors did you focus on today, and what was the most surprising insight you gained?

How helpful was the AI in identifying content gaps and providing inspiration for unique content ideas?

Which competitor strengths or weaknesses are most relevant to your brand strategy?

How comfortable do you feel using the AI to conduct competitor analysis?

What other competitor insights or industry trends would you like to explore with AI?

DAY 14

Developing AI-Driven Audience Engagement Strategies

Goal:

Learn how to use AI to create and implement audience engagement strategies that encourage interaction, build relationships, and increase loyalty. Today's activities will guide you in generating interactive content, crafting engaging responses, and planning personalized outreach. By the end of today, you'll have strategies to strengthen your brand's connection with its audience and foster community around your content.

Activities for Day 14

1. Start a New Thread and Name It

Open a new thread in ChatGPT and name it **"Audience Engagement Practice."** This will keep your prompts and outputs organized as you experiment with engagement strategies.

> **Reminder:** If you haven't set up a Custom GPT yet, start each new thread by copying your **Training Document, Brand Voice Guidelines, and Brand Persona** from your AI Resource Center into the prompt box. This will provide the AI with context, ensuring your brand's tone, style, and audience are accurately represented throughout today's activities.

2. Define Your Audience's Engagement Preferences

Consider how your audience prefers to interact with your brand—whether through social media comments, email responses, or interactive content. If you're unsure, prompt the AI to suggest likely preferences based on your industry and audience type.

Sample Prompt: "You are a community engagement expert helping us understand the preferred engagement methods for eco-conscious consumers. Suggest common ways this audience likes to interact with brands like ours. The tone should be knowledgeable and audience-focused."

Write down 2–3 engagement methods that resonate with your audience:

____

____

____

____

3. Create Polls, Quizzes, or Interactive Posts for Social Media

Interactive posts are a great way to foster engagement and encourage participation. Prompt the AI to create a few interactive ideas, like polls, quizzes, or quick Q&A prompts, based on popular topics for your audience.

Sample Prompt: "You are a social media strategist generating interactive post ideas for eco-conscious consumers on Instagram. Suggest three interactive post ideas that encourage followers to share their sustainability habits. The tone should be engaging and fun."

Choose the most relevant ideas and schedule them into your content calendar for increased social media interaction.

4. Plan Personalized Response Prompts for Social Media Comments

Responding directly to followers' comments or questions can make them feel valued and increase loyalty. Ask the AI to suggest ways to personalize responses for common types of comments, such as questions, compliments, or suggestions.

Sample Prompt: "You are a customer engagement expert creating personalized responses to common social media comments from customers interested in eco-friendly products. Suggest friendly and informative responses to questions about our products' eco-friendliness. The tone should be friendly and appreciative."

Review and save these response templates in your AI Resource Center for easy access.

5. Write Engaging Follow-Up Questions

Follow-up questions can keep conversations going and show genuine interest in your audience's views. Prompt the AI to suggest follow-up questions based on a recent post or campaign topic.

Sample Prompt: "You are a community manager creating follow-up questions for eco-conscious consumers who comment on our post about sustainable kitchen habits. Suggest three questions to encourage deeper conversation and engagement. The tone should be curious and engaging."

Add these questions to your response plan to encourage deeper engagement and foster community.

6. Plan Engagement Content for Stories or Live Q&A Sessions

Consider using stories or live Q&A sessions to provide an interactive experience for your audience. Prompt the AI to suggest story ideas or topics for live Q&A sessions that relate to your brand's mission and audience interests.

Sample Prompt: "You are an engagement strategist developing story and Q&A session ideas for a sustainable home goods brand. Suggest three topics for an Instagram live Q&A related to eco-friendly kitchen practices. The tone should be welcoming and conversational."

Add these engagement sessions to your content calendar and save any outline ideas.

7. Create Educational "How-To" Posts or Guides

Educational posts that teach something valuable can drive engagement by encouraging comments, shares, and saves. Prompt the AI to help you develop "how-to" posts or guides related to popular topics within your industry.

Sample Prompt: "You are an educator in sustainable living creating 'how-to' content for eco-conscious consumers who want to adopt sustainable kitchen practices. Suggest three 'how-to' post ideas that align with our brand voice. The tone should be informative, encouraging, and actionable."

Organize the ideas in your content calendar or save them in your AI Resource Center for easy access.

8. Plan Content to Highlight Customer Stories or Testimonials

Showcasing customer stories or testimonials can build community and make your audience feel valued. Prompt the AI to help structure a post or story series featuring customer feedback or testimonials.

Sample Prompt: "You are a content creator developing a post series to highlight customer testimonials on the impact of our eco-friendly products. Create a plan for a social media series that features different customer stories. The tone should be appreciative and community-focused."

Organize these posts in your content calendar and save the series outline for reference.

9. Create "Thank-You" or Appreciation Messages for Loyal Customers

Showing appreciation for loyal customers can deepen relationships and encourage continued support. Ask the AI to generate messages for loyal customers that convey gratitude and encourage engagement.

Sample Prompt: "You are a brand representative crafting thank-you messages for loyal customers of our eco-friendly product line. Write personalized messages that convey gratitude and encourage customers to share their experiences on social media. The tone should be warm and appreciative." Save these thank-you messages to use across multiple platforms.

Common Mistakes to Avoid

1. Not Defining Clear Engagement Goals

Mistake: Starting engagement efforts without specific goals, which can lead to unstructured and inconsistent interactions.

Better: Set clear goals for engagement, such as building brand loyalty, gathering feedback, or encouraging user-generated content. Defined goals make engagement efforts more focused and impactful.

2. Using Generic Engagement Prompts

Mistake: Relying on broad or overly general engagement questions (e.g., "What do you think?"), which often results in low interaction.

Better: Use targeted, specific prompts that encourage thoughtful responses, such as "What's your favorite eco-friendly kitchen swap?" Specificity encourages deeper engagement and more meaningful conversations.

3. Overlooking Audience Preferences for Interaction Styles

Mistake: Assuming all audiences engage in the same way, which can reduce interaction on certain platforms.

Better: Tailor engagement tactics to your audience's preferred platforms and interaction styles. For instance, use polls and Q&As on Instagram, while creating more discussion-based posts on LinkedIn.

4. Neglecting Follow-Up Responses

Mistake: Posting engagement content without following up on responses, which can make interactions feel one-sided or insincere.

Better: Plan for follow-up comments or questions to deepen the conversation. Engaging further shows that you value your audience's input and helps build community around your brand.

5. Not Personalizing Responses

Mistake: Using standard, automated responses for all comments, which can make engagement feel generic and impersonal.

Better: Add personalized elements to responses, such as using the commenter's name or acknowledging specific feedback. Personalized responses make interactions feel authentic and increase loyalty.

6. Missing Opportunities for Interactive Content

Mistake: Limiting engagement tactics to text-only questions, missing out on more interactive content formats.

Better: Experiment with interactive formats like polls, quizzes, live Q&A sessions, or user-generated content prompts. These formats often lead to higher engagement and can be tailored to fit platform norms.

Review the engagement content you've created today. Are the prompts specific and inviting, and do they align with your engagement goals? Identify one opportunity to add follow-up questions or personalize responses to boost interaction. Document any effective engagement prompts in your AI Resource Center for future use, along with notes on which formats worked best for each platform.

Reflection for Day 14

What audience engagement methods did you focus on today, and which felt most aligned with your brand?

How effective was the AI in generating ideas for interactive content and personalized responses?

Which engagement strategies (e.g., polls, live Q&A, thank-you messages) resonated most with your audience preferences?

How comfortable do you feel using the AI to develop engagement strategies?

What other engagement techniques or content formats would you like to explore with AI?

WEEK 2 RECAP

SYSTEMATIZING YOUR AI WORKFLOW

Congratulations on completing Week 2! You've made significant progress in developing systematic approaches to content creation with AI. Let's review your achievements and reflect on your growth.

Key Skills and Accomplishments

Brand Voice Mastery

- Created a comprehensive **Brand Persona** document
- Developed clear tone and style guidelines
- Built templates for consistent messaging
- Learned to guide AI in maintaining brand voice

Content Strategy Development

- Established clear **Content Pillars**
- Created audience-specific messaging frameworks
- Built reusable content templates
- Developed cross-platform content strategies

Workflow Enhancement

- Set up efficient AI workspace organization
- Created standardized prompting templates

- Developed quality control processes
- Built scalable content creation systems

Reflection on Progress

- How has your AI workflow evolved?
- How has your content creation process changed with these new systems in place?
- Which templates or frameworks have been most helpful?
- What improvements have you noticed in brand consistency?
- How has your efficiency changed with these new processes?

Common Challenges and Solutions

The transition from creating individual pieces to developing systematic approaches can feel overwhelming at first. You might have struggled to:

- Define your brand voice precisely
- Decide which templates to create

These challenges are part of building a robust content system. The key is to start with core templates and gradually expand them based on your needs. Remember that your systems will evolve as you use them more frequently and identify areas for improvement.

Looking Ahead

As we move into Week 3, you'll apply these systems to more advanced marketing initiatives. The frameworks and templates you've created will serve as the foundation for sophisticated campaigns and targeted content strategies. You'll see how your investment in systematic approaches pays off when tackling complex marketing challenges.

Take pride in the systems you've built this week. You've transformed from someone who can use AI effectively to someone who can leverage AI systematically for consistent, high-quality content creation. This is a crucial step in becoming a proficient AI marketer!

WEEK 3 OVERVIEW

CREATING PROFESSIONAL MARKETING MATERIALS WITH AI

This week focuses on developing more substantial marketing materials that showcase your brand's expertise. You'll learn to use AI to create professional documents like case studies and white papers, while also exploring ways to reach and engage your audience more effectively.

What You'll Learn This Week

1. Long-Form Content Creation
   - Writing compelling case studies
   - Developing informative white papers
   - Crafting professional presentations
   - Repurposing content across formats
2. Media and PR Development
   - Creating press releases
   - Developing media pitches
   - Writing guest blog proposals
   - Building PR outreach materials
3. Audience Understanding
   - Analyzing customer feedback
   - Creating targeted content

- Developing personalized messaging
- Understanding audience segments

Moving Beyond Basics

This week bridges the gap between basic content creation and more sophisticated marketing materials. You'll develop confidence in handling longer, more complex pieces while maintaining quality and brand consistency. Focus on understanding how these materials fit into your broader marketing strategy.

DAY 15

Refining Brand Tone and Style Consistency Across Platforms & Creating Social Media Profiles

Goal:

Learn how to refine and apply your brand's tone and style consistently across different social media platforms, while setting up professional profiles that reflect your brand identity. By the end of today, you'll have a clear approach to maintaining brand consistency and polished social profiles on LinkedIn, Facebook, Instagram, and X.

Activities for Day 15

1. Start a New Thread and Name It

Open a new thread in ChatGPT and name it "Brand Tone and Social Media Profiles." This will keep your prompts and outputs organized as you work on refining tone and setting up profiles.

Reminder: If you haven't set up a Custom GPT yet, start each new thread by copying your Training Document, Brand Voice Guidelines, and Brand Persona from your AI Resource Center into the prompt box. This will provide the AI with context, ensuring your brand's tone, style, and audience are accurately represented throughout today's activities.

2. Refine Brand Tone and Style for Each Platform

Each social media platform has its own norms, so it's helpful to refine your brand tone slightly to fit each while staying consistent overall. Prompt the AI to provide tone guidelines for LinkedIn, Facebook, Instagram, and X.

Sample Prompt:

"You are a brand strategist refining our brand tone and style for each social media platform. Suggest tone guidelines for LinkedIn, Facebook, Instagram, and X, maintaining a consistent voice while adapting to each platform's audience. Focus on friendliness, professionalism, and approachability."

Write down your brand's tone for each platform:

- LinkedIn : ______________________________
- Facebook : ______________________________
- Instagram : ______________________________
- X (Twitter): ______________________________

3. Create Profile Bios Featuring the Founder or Brand Face

To personalize the brand, create bios that introduce the founder, CEO, or public face of the brand. Ask the AI to draft bios that reflect this individual's connection to the brand's mission and values. Upload the most current bio or social profiles before asking the AI to refine them.

Sample Prompt:

"You are a copywriter creating profile bios for LinkedIn, Facebook, Instagram, and X that introduce our founder and highlight their role in our sustainable home goods brand. The LinkedIn bio should be professional and informative, the Facebook bio friendly and approachable, the Instagram bio engaging and concise, and the X bio brief and impactful. Use the attached information to guide you."

Try this: Review the bios and adjust as needed to keep them on-brand. Save these bios in your AI Resource Center under Social Media Profiles.

4. Generate Brand-Aligned Visual Ideas for Each Profile

Visual consistency is key to a cohesive brand presence. Ask the AI to suggest profile and cover image ideas for each platform that reflect your brand identity.

Sample Prompt:

"You are a visual branding expert for a sustainable home goods brand. Suggest profile and cover image ideas for LinkedIn, Facebook, Instagram, and X that convey sustainability, innovation, and warmth."

Try this: Review the suggestions, note any preferred ideas, and save them these in your AI Resource Center under Social Media Profiles for future reference.

5. Plan Platform-Specific Content Themes

Develop a content theme or main focus for each platform. This will help ensure your messaging aligns with audience expectations on each site.

Sample Prompt:

"You are a content strategist creating platform-specific themes for LinkedIn, Facebook, Instagram, and X. Suggest a unique content focus for each platform that aligns with our brand's eco-friendly mission and connects with the platform's audience."

Write down your content themes for each platform:

- LinkedIn : ______________________________
- Facebook : ______________________________
- Instagram : ______________________________
- X (Twitter): ______________________________

6. Customize Your Posting Frequency and Schedule

Each platform benefits from a tailored posting schedule. Prompt the AI to suggest posting frequencies for each platform based on your brand's audience and content themes.

Sample Prompt:

"You are a social media planner recommending posting schedules for LinkedIn, Facebook, Instagram, and X for a sustainable home goods brand. Suggest posting frequencies and optimal days/times to engage with eco-conscious consumers."

Try this: Choose the suggested schedules that best fit your brand's capacity and audience habits, and add them to your AI Resource Center under Social Media Posting Schedules.

7. Craft Your Brand's Unique Hashtag Set for Each Platform

Hashtags can increase reach and engagement when tailored for each platform. Prompt the AI to generate hashtags for LinkedIn, Facebook, Instagram, and X, customized to fit each platform's character limit and audience.

Sample Prompt:

"You are a social media strategist creating platform-specific hashtags for a sustainable home goods brand. Suggest hashtags for LinkedIn, Facebook, Instagram, and X that reflect our focus on eco-friendly products, home sustainability, and conscious consumerism."

Try this: Choose the hashtags that best align with your brand and save them for future reference. Add them to your AI Resource Center under Social Media Hashtags.

8. Create a Consistency Checklist for Social Media Profiles

To ensure future consistency, prompt the AI to create a checklist for maintaining brand tone, visuals, and engagement style across all platforms.

Sample Prompt:

"You are a social media branding expert creating a consistency checklist for LinkedIn, Facebook, Instagram, and X. Suggest items for maintaining a consistent tone, visual identity, and engagement approach across all platforms."

Try this: Save this checklist in your AI Resource Center under Social Media Consistency Guide for regular reference when updating profiles or creating new content.

Common Mistakes to Avoid

1. Inconsistent Tone Across Social Media Profiles

Mistake: Using varying tones on different profiles (e.g., casual on one platform, formal on another), leading to a fragmented brand image.

Better: Establish a unified brand tone that adapts slightly for each platform's audience. Document these tone variations in your brand guidelines to maintain consistency across LinkedIn, Instagram, Facebook, and X.

2. Neglecting Platform-Specific Bio Customization

Mistake: Copying the same bio across all platforms without tailoring it to suit each one's unique audience and character limits.

Better: Customize each bio to fit the platform. For instance, keep LinkedIn bios professional, make Instagram bios more conversational, and keep X bios brief and punchy. Tailored bios help you connect better with each audience.

3. Overlooking Visual Consistency

Mistake: Using random profile or cover images that don't reflect brand identity, which can make profiles feel disconnected.

Better: Choose visuals that align with your brand's color palette, style, and personality. Maintain a consistent look across platforms while making slight adjustments to fit each platform's design requirements.

4. Ignoring Hashtag Strategy

Mistake: Using the same hashtags across all platforms or none at all, which can limit reach and relevance.

Better: Develop a hashtag strategy for each platform. For instance, use a mix of popular and branded hashtags on Instagram, while keeping hashtags minimal or professional on LinkedIn. Tailoring hashtags boosts discoverability and engagement.

5. Forgetting to Document Posting Schedules

Mistake: Posting inconsistently or on platforms without planned timing, which can reduce engagement and follower loyalty.

Better: Establish a regular posting schedule for each platform based on audience behavior, and document it in your AI Resource Center. Consistent scheduling keeps your brand active and predictable for followers.

6. Missing Calls-to-Action in Profiles

Mistake: Not including a clear CTA in bios, which misses opportunities to direct followers to actions like signing up or visiting your website.

Better: Add a CTA that reflects each platform's audience and your main goals, such as "Explore our products" or "Join our newsletter." A well-placed CTA guides followers to the next step in engaging with your brand.

Try this: Review each social media profile. Are the bios customized to fit platform norms, and do they include a clear CTA? Check for visual and tonal consistency, and ensure that hashtags and posting schedules are documented in your AI Resource Center. Refine as needed to maintain a cohesive brand presence across all profiles.

Reflection for Day 15

How well did the AI capture your brand tone across different social media platforms?

Which platform-specific guidelines or themes felt most aligned with your brand's identity?

Which content themes or posting schedules do you feel will be most effective for your audience?

How comfortable do you feel using the AI to create a cohesive brand presence across multiple platforms?

What other social media engagement techniques would you like to explore with AI?

DAY 16

Creating a Referral Program

> **Goal:**
> Learn how to use AI to design an effective referral program that incentivizes your customers to refer others to your brand. Today's activities will help you create a structured program, draft messaging, and plan follow-up communication to encourage referrals. By the end of today, you'll have a clear referral strategy that leverages your loyal customers to expand your reach.

Activities for Day 16

1. Start a New Thread and Name It

Open a new thread in ChatGPT and name it "Referral Program Creation." This will help you keep your prompts and outputs organized as you work on designing your referral program.

> **Reminder:** If you haven't set up a Custom GPT yet, start each new thread by copying your Training Document, Brand Voice Guidelines, and Brand Persona from your AI Resource Center into the prompt box. This will provide the AI with context, ensuring your brand's tone, style, and audience are accurately represented throughout today's activities.

2. Define Your Program Goals and Incentives

Start by defining the main goal of your referral program (e.g., increase customer acquisition, build brand loyalty) and the types of incentives you'll offer. Prompt the AI to suggest incentive ideas, such as discounts, rewards, or loyalty points.

Sample Prompt:

"You are a marketing strategist helping create our referral program for a sustainable home goods brand. Suggest referral incentives that would appeal to eco-conscious customers, like discounts, loyalty points, or special offers."

Write down your program's primary goal and selected incentives:

____

____

____

____

3. Develop the Program's Key Structure and Rules

Ask the AI to outline the basic structure of our referral program, including eligibility requirements, referral limits, and reward thresholds. Having clear rules ensures transparency and encourages participation.

Sample Prompt:

"You are a referral program strategist creating guidelines for our sustainable home goods brand. Outline the main structure of our referral program, including eligibility, referral limits, reward levels, and any conditions customers should know about."

Try this: Adjust the AI's structure as needed, then save it in your AI Resource Center under Referral Program Guidelines.

4. Craft a Referral Invitation Message

Create a message inviting customers to join our referral program. This message can be shared via email, social media, or your website, so it should be engaging and clearly explain the benefits of joining.

Sample Prompt:

"You are a copywriter creating a referral program invitation message for our sustainable home goods brand. Write an engaging message inviting customers to join our referral program, highlighting the rewards they'll earn by referring friends."

Try this: Save the finalized message for future reference in your AI Resource Center under Referral Program Messages.

5. Generate Messaging for Referring Customers to Share

Provide your referring customers with easy-to-share messages they can use to invite friends. Prompt the AI to create short referral messages for social media, email, or text.

Sample Prompt:

"You are a copywriter creating referral messages for customers of our sustainable home goods brand to share with their friends. Write a few brief messages customers can post on social media, send by email, or text to invite friends to check out our brand."

Try this: Choose the best referral messages and save them in a shared location your AI Resource Center for customers to access easily.

6. Draft Follow-Up Messages for Referred Customers

Once someone has been referred, it's helpful to follow up with a message that introduces your brand and explains any referral incentives they'll receive. Ask the AI to create a friendly follow-up message for new referrals.

Sample Prompt:

"You are a customer engagement specialist writing a follow-up message for new customers referred to our sustainable home goods brand. Write a welcoming message that introduces the brand, explains the referral incentive, and invites them to explore our products."

Try this: Save the follow-up message in your AI Resource Center under Referral Program Messages for easy access.

7. Plan Incentive Reminders for Participants

Keep customers engaged by reminding them periodically about their progress in the referral program and any upcoming rewards. Prompt the AI to draft reminders that motivate participants to continue referring friends.

Sample Prompt:

"You are a marketing strategist creating incentive reminder messages for our sustainable home goods brand's referral program. Write friendly reminder messages that encourage participants to refer more friends and update them on their referral progress."

Try this: Schedule these reminders at regular intervals and save the messages in your AI Resource Center under Referral Program Messages for future reference.

8. Create a Tracking and Reward Distribution Plan

Effective referral programs rely on smooth tracking and reward distribution. Prompt the AI to suggest methods for tracking referrals and distributing rewards to participants in a way that's seamless and user-friendly.

Sample Prompt:

"You are a program administrator for our sustainable home goods brand's referral program. Suggest tracking methods for referrals and an easy way to distribute rewards to customers who meet referral milestones."

Try this: Choose the tracking and reward distribution methods that fit best, and note them in your AI Resource Center under Referral Program Operations.

9. Develop a Referral Program FAQ

Provide a set of FAQs that clarify common questions customers may have about our referral program. This helps ensure transparency and improves customer experience.

Sample Prompt:

"You are a customer service specialist creating an FAQ for our sustainable home goods brand's referral program. Write responses to common questions like 'How do I refer friends?' and 'When will I receive my rewards?'"

Try this: Save the FAQ in your AI Resource Center under Referral Program FAQs for quick reference on your website or social media.

Common Mistakes to Avoid

1. Launching a Referral Program Without Clear Incentives

Mistake: Creating a referral program with vague or minimal rewards, which can result in low participation.

Better: Define attractive incentives for both referrers and their referrals, such as discounts, exclusive offers, or loyalty points. Clear, valuable incentives increase engagement and motivate customers to refer others.

2. Ignoring Simplicity in Program Rules

Mistake: Overloading the referral program with complex rules or unclear eligibility requirements, which can deter participation.

Better: Keep rules straightforward and easy to understand. Ensure that customers know how to participate, what rewards they'll receive, and any limitations. A simple, transparent program encourages more referrals.

3. Not Providing Ready-to-Use Referral Messages

Mistake: Expecting customers to write their own referral messages, which can limit engagement.

Better: Create easy-to-share, pre-written referral messages for email, social media, or text. These make it simple for customers to spread the word and can increase the reach of your program.

4. Failing to Follow Up with New Referrals

Mistake: Missing the chance to engage with referred customers immediately, which can result in lower retention.

Better: Send follow-up messages to new referrals, welcoming them to the brand and reinforcing the value they'll find. A friendly introduction can help build a positive impression and encourage loyalty.

5. Not Tracking or Communicating Referral Progress

Mistake: Running a referral program without notifying participants of their progress or earned rewards, which may lead to disengagement.

Better: Provide regular updates on referral status and progress toward rewards. Reminders or "just one more referral to earn..." messages can keep participants motivated and engaged.

6. **Overlooking the Importance of Program Promotion**

Mistake: Relying on a passive approach, like a single email or website banner, to drive referrals.

Better: Actively promote the program through multiple channels, including email newsletters, social media posts, and in-purchase follow-ups. Consistent promotion increases visibility and participation.

Try This: Upload your referral program details, take a screenshot, or input a URL link, and ask the AI to review it. Request feedback on the program's clarity, incentive structure, and ease of participation. Use AI's suggestions to simplify any confusing elements or to refine the incentive messaging for stronger engagement.

Reflection for Day 16

What incentives did you choose for your referral program, and how well do you think they'll resonate with your audience?

Which parts of the program structure felt most useful for keeping the program clear and accessible?

How comfortable do you feel using AI to create messages and reminders for a referral program?

How comfortable do you feel using AI to create messages and reminders for a referral program?

What other aspects of customer-driven marketing would you like to explore with AI?

DAY 17

Offline Marketing Strategy – Event Promotion and Step-by-Step Checklists

Goal:

Learn how to use AI to build a strategic plan for promoting a live event, from developing promotional messaging to creating step-by-step checklists for offline marketing tasks. By the end of today, you'll have a clear strategy and actionable checklists to promote an event effectively in physical spaces.

Activities for Day 17

1. Start a New Thread and Name It

Open a new thread in ChatGPT and name it **"Offline Marketing Strategy Practice".** This will help keep prompts and outputs organized as you work through today's activities.

Reminder: If you haven't set up a Custom GPT yet, start each new thread by copying your Training Document, Brand Voice & Tone Guidelines, and Brand Persona from your AI Resource Center into the prompt box. This will provide the AI with context, ensuring your brand's tone, style, and audience are accurately represented throughout today's activities.

2. Build an Event Promotion Strategy for an Upcoming Event

If you have an upcoming event or campaign, prompt the AI to help you create an event promotion strategy that includes objectives, key messaging, target audience, and timing for offline promotional activities.

Sample Prompt:

"You are an event strategist helping create a promotion plan for an eco-friendly home goods brand's upcoming product launch event. Outline the objectives, key messages, target audience, and recommended timing for offline promotions, such as posters, fliers, and community announcements."

Save the event strategy for easy access and future planning.

3. Develop Key Messaging for Event Announcements

Ask the AI to help you create concise, engaging messaging for offline announcements (e.g., community bulletin boards, local newsletters, store posters). These messages should include a headline, event details, and a CTA.

Sample Prompt:

"You are a copywriter creating offline announcement messaging for our sustainable home goods brand's Earth Day event. Write a compelling headline, brief event description, and a CTA that encourages people to attend."

Write down your favorite messaging for offline announcements:

____

____

____

____

4. Create a Checklist for Pre-Event Preparation Tasks

A detailed pre-event checklist ensures all necessary tasks are completed before the event. Prompt the AI to develop a pre-event checklist that includes tasks like venue booking, promotional materials, and local advertising, formatted as a checklist with boxes and no bullet points.

Sample Prompt:

"You are an event coordinator creating a pre-event checklist for our sustainable home goods brand's product launch event. List essential tasks, such as booking the venue, coordinating printed materials, and arranging local advertising, and format it as a checklist with boxes and no bullet points."

Save this checklist for easy access and future use.

5. Draft a Timeline for Event-Day Tasks in Table Format

Creating a timeline for the day of the event can help you coordinate activities and ensure smooth execution. Ask the AI to generate a timeline in table format that covers event setup, registration, customer engagement, and teardown.

Sample Prompt:

"You are an event planner creating an event-day timeline for our sustainable home goods brand's Earth Day launch event. Organize it as a table with columns for 'Time,' 'Task,' and 'Assigned Team,' covering setup, registration, engagement activities, and teardown."

Adjust the timeline based on your event's specifics and save it in a shared location your team can access.

6. Plan Follow-Up Messaging for Attendees

Post-event follow-up helps maintain engagement with attendees and can turn them into loyal customers. Prompt the AI to draft a follow-up message thanking attendees, highlighting key moments, and inviting them to learn more about your brand.

Sample Prompt:

"You are a customer engagement specialist creating a follow-up message for attendees of our Earth Day event. Write a thank-you message that showcases event highlights and invites them to explore our eco-friendly products online."

Save the follow-up message for reference after the event.

7. Create a Checklist for Post-Event Wrap-Up Tasks

A post-event checklist helps ensure that you complete all necessary follow-up tasks, from gathering feedback to evaluating event success. Ask the AI to develop a post-event checklist that covers these tasks, formatted as a checklist with boxes and no bullet points.

Sample Prompt:

"You are an event coordinator creating a post-event checklist for our sustainable home goods brand's product launch event. Include tasks like sending follow-up emails, gathering attendee feedback, and assessing event outcomes. Format it as a checklist with boxes and no bullet points."

Save the post-event checklist for future events.

Common Mistakes to Avoid

1. Not Setting Clear Event Promotion Goals

Mistake: Starting event promotion without defining specific goals, which can lead to unfocused messaging and ineffective tactics.

Better: Identify measurable goals for the event, such as increasing attendance by a certain percentage or generating specific types of leads. Clear goals help shape a focused promotion strategy.

2. Overloading Messages with Excessive Details

Mistake: Including too many details in announcements, which can overwhelm readers and make key information easy to miss.

Better: Prioritize essential information—like date, time, location, and the primary benefit of attending. Provide additional details through follow-up messages or on the event landing page.

3. Failing to Adapt Messaging for Offline and Online Channels

Mistake: Using the same message for both offline and online promotions, which may not suit each channel's format or audience.

Better: Tailor your messaging to fit each platform. For instance, offline fliers should be highly visual with concise text, while emails can include more detailed information and links.

4. Neglecting to Schedule Pre- and Post-Event Follow-Ups

Mistake: Only sending the initial announcement without scheduling reminders or follow-ups, which can reduce attendance and engagement.

Better: Set up a series of reminders leading up to the event and post-event follow-ups thanking attendees or providing additional resources. Regular touchpoints keep your audience engaged and informed.

5. Overlooking Cross-Promotion Opportunities

Mistake: Focusing only on your existing channels and audience, missing the chance to expand reach through partnerships.

Better: Explore partnerships with complementary brands, local influencers, or community organizations to extend your reach. Cross-promotion can attract new attendees who might not otherwise see your event.

6. Forgetting to Use Visual and Interactive Elements

Mistake: Relying solely on text-based promotion, which may not capture attention as effectively.

Better: Incorporate engaging visuals like event graphics, countdown timers, or teaser videos to generate excitement. Visuals make promotions more eye-catching, especially for social media and offline materials.

Upload your event materials, take a screenshot, or provide a URL for AI review. Ask for feedback on the clarity, messaging, and visual appeal of your promotional assets. Use AI's suggestions to refine any areas, such as simplifying messaging, improving visuals, or creating a reminder schedule, to strengthen engagement.

Reflection for Day 17

Which part of the event promotion strategy felt most effective for engaging your target audience?

What messaging did you find most compelling for offline announcements?

How helpful were the checklists and timelines for planning the event?

How comfortable do you feel using AI to support your event promotion strategies?

What other offline marketing strategies would you like to experiment with using AI?

DAY 18

Developing Paid Advertising Copy for Social Media and Search Engines

Goal:

Learn how to create effective ad copy for paid advertising on social media and search engines that captures attention, highlights key benefits, and drives clicks. By the end of today, you'll have tailored ad copy that aligns with your brand and motivates your audience to take action.

Activities for Day 18

1. Start a New Thread and Name It

Open a new thread in ChatGPT and name it **"Paid Advertising Copy Practice".** This will help keep prompts and outputs organized as you work on today's activities.

Reminder: If you haven't set up a Custom GPT yet, start each new thread by copying your Training Document, Brand Voice & Tone Guidelines, and Brand Persona from your AI Resource Center into the prompt box. This will provide the AI with context, ensuring your brand's tone, style, and audience are accurately represented throughout today's activities.

2. Upload Previous Ad Examples or Inspiration

If you have previous ads from past campaigns, create a document with your top-performing ads and upload it into the prompt. This will allow the AI to learn your brand's advertising style and mimic it effectively. If you don't have prior ads, you can gather inspiration by finding ads you like and taking screenshots. Add these to a document with notes about why they resonate with you.

Upload Prompt:

"You are an ad copywriter reviewing our top-performing ads [or preferred ad examples] to learn our advertising style. Analyze these ads and provide a summary of their style, tone, and key messaging elements."

Save the AI's summary for future ad creation.

3. Generate Messaging Ideas for Social Media Ads

Begin by asking the AI to generate a list of messaging ideas for social media ads. These should reflect your brand's tone and highlight key benefits that will capture your audience's attention.

Sample Prompt:

"You are an ad copywriter developing social media ad messages for our sustainable home goods brand. Suggest five different ad messages that appeal to eco-conscious customers, highlighting benefits such as sustainability, durability, and eco-friendly materials."

Choose the most compelling ideas and save them.

4. Create Headline Variations for Search Ads

Headlines in search ads need to be concise and impactful. Prompt the AI to create a variety of headline options that include keywords, capture attention, and align with your brand.

Sample Prompt:

"You are an ad specialist creating search ad headlines for our eco-friendly home goods brand. Write five headline variations that are under 30 characters and highlight keywords like 'eco-friendly,' 'sustainable,' and 'home products.'"

Write down your favorite headline options:

5. Develop Descriptions for Search Ads

Ask the AI to generate short, engaging descriptions for search ads, making sure each one includes a clear CTA and key product benefits.

Sample Prompt:

"You are an ad copywriter creating descriptions for search ads promoting our eco-friendly kitchen products. Write three descriptions under 90 characters that highlight sustainability benefits and include a CTA."

Save the descriptions for easy access during ad setup.

6. Draft Compelling Calls to Action for Ads

Effective CTAs drive conversions by encouraging immediate action. Prompt the AI to suggest CTAs that are relevant for both social media and search ads, focusing on clarity and motivation.

Sample Prompt:

"You are a digital marketer creating CTAs for our eco-friendly home goods brand. Suggest five different CTAs that encourage customers to learn more, make a purchase, or visit our website."

Write down your favorite CTAs:

____

____

____

____

7. Create Audience-Specific Ad Copy for Targeted Campaigns

If you're running ads for different audience segments, ask the AI to customize ad copy based on specific audience interests or demographics.

Sample Prompt:

"You are an ad strategist creating tailored ad copy for our sustainable home goods brand. Write ad messages for two audience segments: (1) eco-conscious young adults, and (2) corporations interested in sustainable office supplies."

Save these segmented ad copies.

8. Design a Checklist for Testing and Refining Ads

To optimize performance, ads should go through a regular testing and refinement process. Prompt the AI to create a checklist, formatted with boxes and no bullet points, to guide you through A/B testing, performance monitoring, and adjustments.

Sample Prompt:

"You are an ad specialist creating a testing and optimization checklist for our paid advertising campaigns. Include steps like A/B testing, monitoring click-through rates, adjusting ad copy, and testing new CTAs. Format it as a checklist with boxes and no bullet points."

Save this checklist in your **AI Resource Center** for future reference.

9. Set Up an Ad Library to Track and Organize Ad Copy

To keep track of all your ads, ask the AI to suggest a format for an ad library. This library can help you monitor ad performance, categorize ads by platform, audience, or campaign, and keep all ad copy organized in one place.

Sample Prompt:

"You are a content manager setting up an ad library for our paid advertising campaigns. Suggest a format for an ad library that includes columns for ad platform, audience segment, campaign name, ad copy, and performance metrics."

Set up the ad library in your preferred software, such as Google Sheets, and save it for easy reference.

Common Mistakes to Avoid

1. Writing Generic Ad Copy Without a Specific Audience in Mind

Mistake: Creating ad copy that speaks to a broad audience, which often results in generic messaging that fails to resonate.

Better: Define a specific target audience for each ad (e.g., eco-conscious consumers or busy professionals) and tailor the language, benefits, and tone to appeal directly to that group.

2. Skipping A/B Testing for Headlines and CTAs

Mistake: Relying on a single headline or CTA without testing alternatives, which can limit ad performance.

Better: Set up A/B tests to compare different headlines, CTAs, and ad variations. Testing allows you to identify the language that drives the most clicks and conversions for your audience.

3. Overlooking the Importance of Platform-Specific Copy

Mistake: Using the same ad copy across all platforms, which can reduce effectiveness due to different norms and audience expectations.

Better: Adjust copy to fit each platform's tone and format. For instance, keep LinkedIn ads professional, make Facebook ads conversational, and ensure Instagram ads are visually driven.

4. Focusing on Features Over Benefits

Mistake: Highlighting product features rather than the benefits they offer, which can make ads feel impersonal or overly technical.

Better: Emphasize the real-world impact of your product or service, like how it saves time, improves sustainability, or simplifies a task. Benefits resonate more with audiences and drive stronger engagement.

5. Using Vague or Weak Calls-to-Action

Mistake: Including vague CTAs like "Learn More" without a clear call to action, which may reduce conversions.

Better: Use strong, specific CTAs like "Start Your Free Trial" or "Discover Sustainable Solutions." Tailored CTAs guide the audience more effectively toward taking the desired action.

6. Not Reviewing AI-Generated Ad Copy for Consistency

Mistake: Copying AI-generated ad suggestions directly without checking for brand tone and consistency, which can result in off-brand messaging.

Better: Review and refine AI-generated ad copy to ensure it aligns with your brand's tone, values, and specific messaging. Small adjustments help keep ads on-brand and improve engagement.

Upload your ad copy examples, take a screenshot, or provide a URL link for AI to evaluate. Request feedback on the clarity, audience specificity, and effectiveness of headlines, CTAs, and benefits highlighted. Use AI's recommendations to refine ad copy for a more targeted, impactful campaign.

Reflection for Day 18

What ad messaging ideas resonated most with your brand and audience?

Which headlines or descriptions did you find most effective for capturing attention?

How comfortable do you feel using AI to develop targeted ad copy?

What other paid advertising techniques or strategies would you like to explore with AI?

DAY 19

Writing for Different Ad Types (Display, Video, Text) and A/B Testing Ideas

Goal:

Learn how to adapt your ad copy for different ad formats, such as display ads, video ads, and text ads. By the end of today, you'll have a set of ad variations that can be used across multiple formats, along with a testing plan to refine your ads based on performance.

Activities for Day 19

1. Start a New Thread and Name It

Open a new thread in ChatGPT and name it **"Multi-Format Ad Copy Practice".** This will help keep prompts and outputs organized as you work on today's activities.

Reminder: If you haven't set up a Custom GPT yet, start each new thread by copying your Training Document, Brand Voice & Tone Guidelines, and Brand Persona from your AI Resource Center into the prompt box. This will provide the AI with context, ensuring your brand's tone, style, and audience are accurately represented throughout today's activities.

2. Upload Previous Ads for Improvement Suggestions

If you have previous ads from past campaigns, upload the top-performing ones into the prompt to get suggestions on how to refine and improve them.

Sample Prompt:

"You are an ad strategist reviewing our past ad campaigns for improvement. Here are our previous ads [upload ads]. Please provide suggestions for enhancing the headlines, CTAs, and overall messaging to increase engagement."

Save the AI's suggestions for reference in future campaigns.

3. Write Ad Copy for Display Ads

Display ads are highly visual, so the copy should be brief but impactful. Ask the AI to create short headlines and descriptions specifically for display ads, ensuring the messaging aligns with the visuals you plan to use.

Sample Prompt:

"You are a display ad copywriter creating short, eye-catching headlines and descriptions for our eco-friendly home goods brand. Write three different headline options and three short descriptions (under 50 characters) for display ads."

Save the chosen display ad copy variations in an organized spreadsheet so you can keep track of them.

4. Develop Script Ideas and B-Roll Suggestions for Video Ads

Video ads allow for storytelling, so focus on messages that engage viewers in a brief but compelling way. Prompt the AI to create a script for a short video ad and suggest B-roll ideas that could enhance the storytelling.

Sample Prompt:

"You are a video ad copywriter creating a 15-second script for a sustainable home goods brand. Write a script that tells a brief brand story, highlights product benefits, and ends with a CTA. Also, suggest B-roll ideas that would enhance the ad, like visuals of eco-friendly materials or product use in daily life."

Save the video ad script and B-roll suggestions and adjust them as needed to fit different video lengths.

5. Craft Ad Copy for Short and Long-Form Text Ads

Text ads are typically seen on search engines and need to be concise and keyword-focused. Ask the AI to create short-form ad copy (under 90 characters) as well as a longer version (up to 200 characters) that includes keywords, main benefits, and a CTA.

Sample Prompt:

"You are a copywriter creating text ad copy for our eco-friendly kitchen products. Write three short ad variations (under 90 characters) and a longer version (under 200 characters) that include keywords like 'sustainable' and 'eco-friendly.'"

Refine and save your preferred short and long-form text ad copy.

6. Create Multiple Variations of Each Ad Type for A/B Testing

To see which ad elements perform best, create several variations of each ad type. Ask the AI to generate different versions by modifying the headlines, CTAs, or tone.

Sample Prompt:

"You are an ad specialist creating variations for A/B testing. Write three alternative versions of each ad type (display, video, text) with different headlines, CTAs, or tones for testing purposes."

Choose the most effective variations and save them for testing.

7. Design a Testing Plan for Multi-Format Ads

Create a structured plan for A/B testing each ad type. Prompt the AI to develop a step-by-step plan that includes setting test objectives, determining sample sizes, and scheduling the tests.

Sample Prompt:

"You are an ad strategist creating a testing plan for multi-format ads (display, video, and text) for our sustainable home goods brand. Outline a step-by-step plan for A/B testing, including test objectives, sample size recommendations, and timing."

Save this plan in your AI Resource Center for ongoing reference.

8. Create a Checklist for Analyzing Ad Performance

A structured checklist for analyzing ad performance will help you understand which elements resonate most with your audience. Ask the AI to create a checklist, formatted with boxes and no bullet points, that includes steps like reviewing click-through rates, assessing engagement, and gathering insights for future improvements.

Sample Prompt:

"You are a marketing analyst creating a checklist for analyzing ad performance across display, video, and text formats. Include steps such as reviewing click-through rates, assessing engagement, and collecting insights for improvement. Format it as a checklist with boxes and no bullet points."

Save this checklist for use after each campaign.

9. Track A/B Test Results and Insights in the Ad Library

To keep track of test results, upload the sample ad library created in Day 18 as context. Then prompt the AI to suggest additional columns or categories to document A/B test results and insights, helping you monitor ad performance over time.

Sample Prompt:

"You are a data analyst updating our ad library to track A/B test results and insights. Here is our sample ad library [paste ad library from Day 18]. Suggest additional columns or categories to include, such as test version, performance metrics, and audience response."

Update your ad library with these additional fields and save it for ongoing use.

Common Mistakes to Avoid

1. Not Adapting Copy for Different Ad Formats

Mistake: Using the same ad copy across display, video, and text ads, which can make ads feel out of place or ineffective in certain formats.

Better: Tailor the language for each format—use concise, visual language for display ads, engaging storytelling for video ads, and keyword-rich text for search ads. AI can assist by generating format-specific variations, making it easier to align messaging with each ad type.

2. Skipping A/B Testing for Multi-Format Ads

Mistake: Running ads in multiple formats without testing variations, which can lead to missed insights on which messages work best.

Better: Use AI to create several versions of each ad, changing elements like the headline, tone, or CTA. Implement A/B testing across formats to identify the most effective combinations and refine future campaigns accordingly.

3. Writing Text-Heavy Display Ads

Mistake: Overloading display ads with too much text, which can overwhelm viewers and reduce click-through rates.

Better: Keep display ad copy minimal and impactful. AI can help by condensing complex ideas into concise, punchy phrases that capture attention in a limited space, while keeping the messaging clear.

4. Using Long Intros in Video Ads

Mistake: Starting video ads with lengthy intros or slow builds, which can cause viewers to lose interest quickly.

Better: Open with a hook or a key benefit in the first few seconds. Use AI to generate quick, engaging intros for video ads that draw viewers in immediately and hold their attention.

5. Not Adding Data-Driven Keywords to Search Ads

Mistake: Leaving out high-impact keywords in search ads, which can reduce discoverability and relevance.

Better: Use AI to research and suggest SEO-friendly keywords for your specific audience and industry. Including these keywords in search ad headlines and descriptions helps improve ad placement and relevance.

6. Failing to Track and Document Test Results

Mistake: Running A/B tests without properly tracking results, which makes it difficult to know which elements are working.

Better: Set up a system to document test results for each ad variation. AI can help you organize these insights by creating a testing report format or by analyzing which ad versions show the highest performance, allowing for more data-driven optimizations.

Upload your multi-format ad examples, screenshots, or a campaign link for AI review. Ask AI to assess each format's suitability, suggest format-specific adjustments, or identify any weak points in the headlines or CTAs. Use AI's recommendations to fine-tune each ad type for stronger engagement and conversion potential across formats.

Reflection for Day 19

Which ad formats (display, video, or text) felt most effective for your brand and message?

How did you find the process of creating variations for A/B testing?

What insights did the A/B testing plan and performance checklist provide for future campaigns?

How comfortable do you feel using AI to develop and test multi-format ad campaigns?

DAY 20

Creating and Promoting a Case Study

Goal:

Today, you'll use AI to create a complete case study based on an interview transcript from a happy client. This case study will demonstrate your brand's impact and provide valuable proof of success to attract new clients. By the end of today, you'll have a promotional asset ready to share across your marketing channels.

Activities for Day 20

1. Start a New Thread and Name It

Open a new thread in ChatGPT and name it **"Case Study Creation Practice".** This will help keep prompts and outputs organized as you work through today's activities.

Reminder: If you haven't set up a Custom GPT yet, start each new thread by copying your Training Document, Brand Voice & Tone Guidelines, and Brand Persona from your AI Resource Center into the prompt box. This will provide the AI with context, ensuring your brand's tone, style, and audience are accurately represented throughout today's activities.

Note: For today's activities, we'll be working with a hypothetical client and interview for demonstration. However, if you have access to a real client story, using actual details and feedback will make the case study even more authentic and effective.

2. Write a Request Email for Case Study Permission

Start by writing an email to request permission to create a case study from one of your client's success stories. Use AI to draft a professional, concise message that explains the purpose and benefits of the case study.

Sample Prompt:

"You are a client relations manager writing an email to request permission to create a case study about a client's experience with our eco-friendly home goods. Write a professional email explaining the purpose and potential benefits of the case study, and ask if they'd be open to a brief interview."

Save the email draft for future use.

3. Write Interview Questions for a Short Testimonial and Longer Case Study

Once permission is granted, develop interview questions that could be used for both a short testimonial and a more detailed case study. Use AI to generate a list of questions that explore the client's experience, challenges, results, and satisfaction.

Sample Prompt:

"You are a content strategist creating interview questions for a client case study about our sustainable home goods brand. Write questions that cover initial challenges, the client's experience with our products, and the results they've seen."

Write down key interview questions:

____

____

____

____

4. Upload the Interview Transcript for AI to Summarize

After conducting the interview, upload the interview transcript to AI and ask it to summarize the main points in a short, one-paragraph case study. This quick summary will help with creating testimonials or teasers for the full case study.

Sample Prompt:

"You are a case study writer summarizing a client interview about their experience with our eco-friendly products. Here is the interview transcript [paste transcript]. Write a concise, one-paragraph summary capturing the main points and results."

Save the summary where you can find it if you need it.

5. Create an Outline and Draft a Full Case Study

Now, use the AI to create an outline and draft a more detailed case study based on the interview transcript. The full case study should include sections like client background, challenges, solution, results, and conclusion.

Sample Prompt:

"You are a content strategist creating an outline and draft for a full case study based on this client interview [paste transcript]. Include sections for the client's background, challenges, our solution, the results, and a conclusion with a CTA."

Save the full draft for client approval.

6. Generate Multiple Testimonial and Case Study Versions

Ask AI to create multiple versions of testimonials and short case studies for use across marketing materials. Generate five short testimonials, three mini case studies, and ten impactful one-line snippets directly based on the client's interview responses.

Sample Prompt:

"You are a content marketer creating promotional content from a client interview. Based on this interview transcript [paste transcript], write five short testimonials, three brief case studies, and ten one-line snippets. Each version should be impactful and highlight different parts of the client's experience."

Revise and save these testimonials, short case studies, and snippets.

7. Send Full Case Study and Testimonials to Client for Approval

Once the full case study and testimonials are drafted, prepare an email to send to the client for their review and approval. Use AI to draft a professional message that includes an invitation for feedback and a clear process for making edits.

Sample Prompt:

"You are a client relations manager sending a case study and testimonials to a client for approval. Write an email to share the draft, ask for their feedback, and outline the next steps if they have any edits or suggestions."

Revise and save the email.

8. Brainstorm Ways to Promote the Case Study and Generate Leads

Finally, ask the AI to suggest promotional strategies for using the case study to generate leads and attract new clients. These strategies could include website placement, social media posts, and incorporating the case study in email campaigns.

Sample Prompt:

"You are a content marketer brainstorming ways to promote a case study and use it to generate leads. Suggest strategies for sharing the case study across our website, social media, and email channels, and ideas for incorporating it into our lead generation efforts."

Save these promotional ideas for future reference.

Common Mistakes to Avoid

1. Skipping the Planning Phase for Case Study Structure

Mistake: Diving into writing without a clear structure, which can lead to a disorganized or unfocused case study.

Better: Outline the case study first, including sections like the client's background, the problem, your solution, and measurable results. AI can assist by generating an outline or suggesting case study frameworks based on your goals and target audience.

2. Relying on Generic or Vague Descriptions

Mistake: Using broad statements or generic language that doesn't convey specific challenges or successes, which can make the case study feel shallow.

Better: Use precise language that clearly describes the client's problem, your solution, and the tangible impact. AI can help by refining vague language or suggesting specific phrasing that adds clarity and depth to each section.

3. Not Highlighting Measurable Results

Mistake: Failing to include quantitative results, which reduces the credibility and impact of the case study.

Better: Include specific metrics, like "40% increase in engagement" or "20% decrease in customer churn," to showcase the effectiveness of your solution. If you have basic results, AI can help by suggesting ways to present data more effectively or by recommending visuals, like charts or graphs, to enhance the impact.

4. Overlooking a Compelling Headline and Summary

Mistake: Using a generic title and intro, which can fail to capture the reader's attention and communicate the value of the case study.

Better: Create a strong headline and brief summary that highlight the core success or benefit, such as "How [Your Solution] Helped [Client Name] Increase Sales by 50%." AI can help brainstorm impactful headlines and summaries that convey the value upfront.

5. Using Passive Language Instead of Active, Results-Oriented Language

Mistake: Writing in a passive voice, which can make the case study feel less dynamic and compelling.

Better: Use active language and results-focused phrasing to convey impact more clearly. For example, say "Our solution reduced costs by 30%," rather than "Costs were reduced by 30%." AI can rewrite passive sentences into active ones, enhancing the clarity and engagement of the narrative.

6. Neglecting to Include a Call to Action

Mistake: Ending the case study without a CTA, which misses the opportunity to guide readers to the next step, like contacting you or exploring related content.

Better: Conclude with a CTA that aligns with the reader's interest, such as "Contact us to see how we can achieve similar results for your business." AI can suggest different CTAs based on your goals for the case study, helping you drive engagement and conversions.

Upload a draft or outline of your case study, take a screenshot, or provide a link for AI review. Ask AI to analyze the structure, clarity of messaging, and effectiveness of the results presented. Use AI's feedback to refine your narrative, adjust metrics presentation, or improve phrasing for greater impact and readability.

Reflection for Day 20

How comfortable do you feel creating structured case studies using AI?

What promotional strategy ideas do you think will be most effective for reaching potential clients?

What other client success stories would you like to explore as potential case studies?

DAY 21

Customer Feedback Collection & Analysis

> **Goal:**
> Today, you'll use AI to gather and analyze customer feedback, helping you identify opportunities for improvement and gain insights into your customers' needs. By the end of today, you'll have a structured approach to collecting, summarizing, and using customer feedback to strengthen your products and services.

Activities for Day 21

1. Start a New Thread and Name It

Open a new thread in ChatGPT and name it **"Customer Feedback Analysis Practice".** This will keep prompts and outputs organized as you work on today's activities.

> **Reminder:** If you haven't set up a Custom GPT yet, start each new thread by copying your Training Document, Brand Voice & Tone Guidelines, and Brand Persona from your AI Resource Center into the prompt box. This will provide the AI with context, ensuring your brand's tone, style, and audience are accurately represented throughout today's activities.

2. Create Customer Feedback Form Questions

Use AI to create a set of thoughtful questions for a short customer feedback survey. Limit the survey to three key questions that cover areas like product satisfaction, service experience, and improvement suggestions.

Sample Prompt:

"You are a customer feedback specialist creating a short feedback form for our sustainable home goods brand. Suggest a mix of open-ended and multiple-choice questions to gather insights about product satisfaction, overall experience, and specific suggestions. Please limit the survey to three impactful questions."

> **Note:** Select the top three questions from the AI's suggestions to keep the survey brief and user-friendly.

3. Write a Follow-Up Email Requesting Feedback

Once the feedback form is ready, draft an email to invite customers to provide their feedback. Use AI to help write a friendly and professional message that emphasizes the value of their input.

Sample Prompt:

"You are a customer relations manager writing a follow-up email requesting feedback from customers who recently purchased our eco-friendly home goods. Write a professional email inviting them to fill out the feedback form and explain how their input will help us improve."

Save the email template for future use.

4. Scrub Data for Personally Identifying Information Before Uploading Responses

> **IMPORTANT**: Before uploading any customer feedback, ensure that you **remove all personally identifying information** from the data. In regulated industries—such as healthcare, finance, and government—take extra care to comply with data privacy and cybersecurity laws in your region. Uploading sensitive data without proper privacy safeguards can lead to legal and security issues.

5. Summarize and Categorize Feedback Responses

Once feedback responses come in, prompt AI to help you analyze the feedback by summarizing key themes. Ask it to categorize feedback into themes, such as product quality, customer service, and product features, to get a clear picture of customer priorities and potential improvements.

Sample Prompt:

"You are a data analyst summarizing feedback from customers who purchased our eco-friendly products. Here are the feedback responses [paste responses]. Organize the feedback into categories like product quality, customer service, and product features, and provide a summary of each category."

Save the summarized feedback for future reference.

6. Generate Key Insights and Actionable Suggestions

Ask AI to help identify specific insights from the feedback and suggest actions based on these insights. For example, if customers mention durability issues, AI might suggest steps for improving product quality or customer instructions.

Sample Prompt:

"You are a customer experience analyst reviewing categorized feedback on our sustainable products. Based on the summary [paste summary], identify three key insights and suggest actionable improvements we can make to address customer concerns."

Refine and save the insights and action suggestions.

7. Draft a Summary Report on Customer Feedback

Compile a summary report that highlights the main findings and includes actionable recommendations. Use AI to draft a report that can be shared with team members, stakeholders, or other departments to improve customer experience.

Sample Prompt:

"You are a customer experience analyst creating a report on recent customer feedback for our sustainable home goods brand. Write a summary report that includes key insights, common themes, and actionable recommendations based on the feedback."

Save the report for ongoing reference.

Common Mistakes to Avoid

1. Asking Vague or Overly Broad Questions

Mistake: Using general questions like "What do you think?" or "How was your experience?" which may lead to unclear or unhelpful responses.

Better: Use specific, targeted questions that encourage detailed feedback, such as "How satisfied were you with our product features?" or "What could we improve in your ordering experience?" AI can help by suggesting precise question formats based on your goals for feedback.

2. Overloading Surveys with Too Many Questions

Mistake: Including too many questions in a survey, which can cause survey fatigue and lead to incomplete responses.

Better: Limit surveys to the most important questions. Prioritize clarity and brevity. AI can assist by helping you trim down your questions or suggest which ones might yield the most valuable insights.

3. Not Segmenting Surveys by Customer Type

Mistake: Sending the same survey to all customers, which can lead to irrelevant answers or missed insights for different groups.

Better: Customize surveys based on customer segments, such as new customers, repeat buyers, or inactive users. AI can help design segment-specific questions to capture insights unique to each group.

4. Failing to Include Open-Ended Questions

Mistake: Relying solely on multiple-choice or rating questions, which may limit feedback depth.

Better: Add a few open-ended questions that allow customers to share specific thoughts in their own words, such as "What features would you like to see added?" AI can help analyze open-ended responses, identifying common themes or suggestions to streamline review.

5. Ignoring Follow-Up on Feedback

Mistake: Collecting feedback without acknowledging or acting on it, which can lead customers to feel unheard or disengaged.

Better: Reach out to thank participants and, where possible, let them know how their feedback has influenced improvements. AI can help draft personalized thank-you messages or even suggest automated follow-ups based on customer responses.

6. Not Using AI to Analyze Trends in Feedback

Mistake: Reviewing each response manually, which can be time-consuming and lead to missed patterns.

Better: Use AI tools to identify trends and common themes across large volumes of feedback. This saves time and helps prioritize actionable insights, such as repeated requests for a specific feature or recurring issues with customer service.

Upload a sample survey, recent feedback, or screenshots of customer responses, and ask AI to review the clarity, relevance, and structure of the questions. AI can also analyze existing feedback to surface trends, helping you quickly spot improvement opportunities or emerging customer needs.

Reflection for Day 21

What feedback insights surprised you, and how could they inform future changes to products or services?

How comfortable do you feel using AI to analyze and interpret customer feedback?

What improvements or strategies would you prioritize based on this feedback?

WEEK 3 RECAP

ADVANCED MARKETING CONTENT WITH AI

Congratulations on completing Week 3! This week, you worked on creating advanced marketing content with AI, focusing on professional-level outputs and versatile content strategies. From polished blog posts to impactful case studies, you've added new tools to your AI skillset. Let's review what you accomplished.

Key Skills and Accomplishments

Blog Post Creation

- Mastered AI-assisted blog content creation by brainstorming topics, developing detailed outlines, and drafting high-quality posts.
- Focused on structuring blogs with engaging introductions, logical flow, and strong CTAs.
- Learned to refine AI outputs for tone, accuracy, and readability, ensuring they align with your audience's needs.

Email Marketing Content

- Created email content for newsletters, announcements, and outreach campaigns.
- Practiced writing subject lines, body content, and CTA that are tailored to your audience and objectives.
- Experimented with AI to match tone and style for different audiences, whether professional, casual, or promotional.

Repurposing Content

- Turned AI-generated blogs into other forms of content, such as email campaigns and social media posts, making your marketing efforts more efficient.
- Used AI to adapt content for multiple platforms while keeping messaging consistent across channels.

Case Study Creation

- Crafted a case study using AI, showcasing a real-world success story with a clear structure: problem, solution, and results.
- Learned how to use AI to highlight key details, create compelling narratives, and align the tone with your brand voice.
- Focused on ensuring that the case study demonstrated value to your audience and aligned with your marketing goals.

Advanced Formatting and Editing

- Polished AI-generated content to meet professional standards, with attention to grammar, tone, and structure.
- Learned to give AI precise instructions for formatting (e.g., headings, bullet points, and paragraph structure) to save time on revisions.
- Enhanced clarity and engagement through iterative editing processes.

Multi-Purpose Content Strategies

- Developed skills for creating versatile content that can be reused across multiple channels and purposes.
- Gained experience in adapting and refining content to suit different platforms while maintaining a cohesive message.

Reflection on Progress

Think back to everything you accomplished this week:

- Which types of content creation felt easiest or most intuitive with AI?
- How has your confidence in refining and formatting AI outputs improved?
- What new insights did you gain about repurposing content or creating case studies?

These reflections can help you identify areas where you're excelling and pinpoint skills to continue developing.

Common Challenges and Solutions

This week may have introduced challenges like creating content that feels polished and professional or adapting outputs for multiple platforms without losing consistency. To tackle these issues, breaking large tasks into smaller steps–such as starting with outlines and refining drafts over time–proved essential.

Another challenge might have been writing case studies that balance detail with storytelling. By using AI to assist with structure and then adding your personal expertise, you created case studies that effectively communicate value.

Looking Ahead

In Week 4, you'll focus on personalized content and automation, learning to create highly specific, audience-targeted materials. You'll also explore ways to save time by automating repetitive tasks, all while maintaining a personal touch in your communications.

Take a moment to celebrate your progress! You've mastered key skills in blogging, email writing, and creating versatile content, including professional-level case studies. These accomplishments will serve as a strong foundation for your next steps. Keep going—you're doing fantastic work!

WEEK 4 OVERVIEW

CREATING YOUR FIRST MARKETING CAMPAIGN

In this final week, you'll apply your AI skills to create product launch materials and your first lead generation funnel. You'll see how your new skills combine to support real marketing goals.

What You'll Learn This Week

1. **Product Launch Content**
   - Writing product launch announcements
   - Creating promotional content
   - Developing social media posts
   - Writing follow-up messages
2. **Lead Generation Introduction**
   - Creating your first lead magnet
   - Writing landing page content
   - Building a welcome email sequence
   - Planning a simple webinar
3. **Content Organization**
   - Keeping marketing materials organized
   - Maintaining consistent messaging
   - Writing clear CTAs
   - Using your AI Resource Center effectively

Practical Application

This week shows you how to combine the individual skills you've learned into basic marketing campaigns. You'll focus on creating clear, consistent messaging that helps achieve specific marketing goals.

DAY 22

Customer Retention Strategy and Revenue Growth

> **Goal:**
> Today, you'll use AI to create a customer retention strategy designed to keep your existing customers engaged and increase their lifetime value. This includes creating personalized follow-up messaging, brainstorming loyalty program ideas, and designing re-engagement email sequences. Additionally, you'll explore revenue-boosting tactics targeted at your existing customer base, such as upselling, cross-selling, and potential subscription models. By the end of today, you'll have a clear plan for both retaining customers and growing revenue through targeted strategies.

Activities for Day 22

1. Start a New Thread and Name It

Open a new thread in ChatGPT and name it **"Customer Retention and Revenue Growth Practice".** This will help keep prompts and outputs organized as you work through today's activities.

> **Reminder:** If you haven't set up a Custom GPT yet, start each new thread by copying your Training Document, Brand Voice & Tone Guidelines, and Brand Persona from your AI Resource Center into the prompt box. This will ensure the AI aligns with your brand's tone, style, and audience throughout today's activities.

2. Write Personalized Follow-Up Messages

Use AI to draft customized follow-up messages that keep your customers engaged post-purchase. Personalized follow-ups can strengthen customer relationships and increase the chances of repeat business.

Sample Prompt:

"You are a customer success manager developing personalized follow-up messages for a sustainable home goods brand. Write follow-up messages for customers one week, one month, and three months after their purchase. Each message should be warm, professional, and subtly encourage re-engagement."

Save the follow-up messages for future reference.

3. Brainstorm Loyalty Program Ideas

Use AI to brainstorm and outline potential loyalty program ideas that reward and incentivize repeat purchases.

Sample Prompt:

"You are a marketing strategist helping a sustainable home goods brand develop loyalty program options. Suggest three different loyalty program structures (e.g., points-based, referral-based, VIP tiers) and explain the potential benefits of each."

Choose the loyalty program ideas that resonate most with your brand's goals, and save them for future use.

4. Create a Re-Engagement Email Sequence

Use AI to craft a series of re-engagement emails for customers who haven't interacted with your brand in a while. Re-engagement emails can help bring back lapsed customers and rekindle their interest.

Sample Prompt:

"You are a content marketer creating a three-part re-engagement email sequence for a sustainable home goods brand targeting inactive customers. Write an email series that reminds them of the brand's value, shares any recent product updates, and includes a special offer or discount to encourage their return."

Refine and save the email sequence.

5. Brainstorm Upselling and Cross-Selling Opportunities

Use AI to identify potential upselling and cross-selling tactics to maximize revenue from existing customers. Focus on ways to add value to their experience without being overly pushy.

Sample Prompt:

"You are a sales strategist developing upselling and cross-selling ideas for a sustainable home goods brand. Based on a typical customer purchase of eco-friendly kitchen products, suggest complementary products or bundles that would add value to the customer experience."

Save your upselling and cross-selling ideas for future reference.

6. Explore Subscription or Membership Model Options

Use AI to brainstorm potential subscription or membership models that could increase recurring revenue from existing customers.

Sample Prompt:

"You are a business development consultant exploring subscription or membership options for a sustainable home goods brand. Suggest a few models (e.g., monthly delivery, members-only discounts, exclusive content access) that would appeal to customers interested in eco-friendly products."

Refine and save the subscription or membership model ideas for future use.

7. Create a Customer Retention and Revenue Growth Plan Summary

Compile your best ideas into a concise customer retention and revenue growth plan. This document should summarize your personalized follow-up messaging, loyalty program ideas, re-engagement strategy, and upselling or subscription models.

Sample Prompt:

"You are a marketing strategist creating a customer retention and revenue growth plan for a sustainable home goods brand. Summarize the key elements of our strategy: personalized follow-ups, loyalty program, re-engagement emails, upselling opportunities, and subscription options."

Save the summary for future reference.

Common Mistakes to Avoid

1. Not Defining Clear Retention Goals

Mistake: Approaching customer retention without specific goals, which can lead to unstructured strategies.

Better: Set measurable retention goals, such as reducing churn rate by a certain percentage or increasing repeat purchases. AI can assist by analyzing historical customer data to identify trends and suggest realistic retention targets.

2. Using Generic Follow-Up Messaging

Mistake: Sending out basic or generic follow-up messages, which can make customers feel undervalued.

Better: Personalize follow-up messages based on customer preferences or purchase history. Use AI to generate tailored messaging for different customer segments, improving relevance and engagement.

3. Overlooking Loyalty Program Customization

Mistake: Implementing a one-size-fits-all loyalty program, which may not resonate with diverse customer needs.

Better: Create tiered loyalty options or unique perks for different customer segments. AI can help brainstorm program structures that align with customer demographics and purchasing patterns.

4. Neglecting Regular Re-Engagement with Lapsed Customers

Mistake: Only reaching out to inactive customers occasionally, which reduces re-engagement chances.

Better: Schedule a re-engagement campaign with multiple touchpoints. AI can generate re-engagement email sequences that remind customers of your brand's value and offer incentives to return.

5. Ignoring Upselling and Cross-Selling Opportunities

Mistake: Missing chances to add value through relevant upsells or cross-sells, which can limit revenue growth.

Better: Use AI to analyze customer purchase history and suggest related products for targeted upselling or cross-selling, enhancing the customer experience while increasing sales.

6. Failing to Document Insights and Actions

Mistake: Not recording insights and action plans for retention strategies, which can lead to inconsistencies in future campaigns.

Better: Document insights in your AI Resource Center, creating a structured reference for customer feedback, retention strategies, and revenue growth opportunities.

Upload your retention strategy, loyalty program outline, or re-engagement email sequence, and ask AI for feedback on clarity, relevance, and engagement potential. Use AI's suggestions to refine your approach, ensuring that it aligns with your audience and retention goals.

Reflection for Day 22

How comfortable do you feel using AI to develop customer retention strategies?

Which loyalty or re-engagement strategy do you think would be most effective for your brand's audience?

What additional revenue opportunities can you identify that would align with your brand and customer base?

DAY 23

Creating Proposals with AI

Goal:

Today, you'll use AI to streamline the process of creating client proposals. You'll draft, refine, and structure a proposal that clearly outlines the project scope, objectives, deliverables, and benefits tailored to an interior designer's needs. By the end of today, you'll have a professional proposal draft ready to customize and send.

Activities for Day 23

1. Start a New Thread and Name It

Open a new thread in ChatGPT and name it **"Proposal Creation Practice".** This will help keep prompts and outputs organized as you work through today's activities.

Reminder: If you haven't set up a Custom GPT yet, start each new thread by copying your Training Document, Brand Voice & Tone Guidelines, and Brand Persona from your AI Resource Center into the prompt box. This will ensure the AI aligns with your brand's tone, style, and audience throughout today's activities.

2. Draft a Proposal Introduction

Use AI to help you craft a compelling introduction for your proposal. This section should briefly introduce your company, establish rapport, and summarize the client's needs or vision for the project.

Sample Prompt:

"You are a client relations manager writing an introductory paragraph for a proposal to an interior designer. The designer is interested in using our eco-friendly home goods to furnish a new residential project. Introduce our company, acknowledge their design vision, and position us as a sustainable solution."

Save the introduction in your AI Resource Center under Proposal Templates.

3. Outline Key Proposal Sections

Ask AI to generate an outline for your proposal, including sections like project scope, objectives, deliverables, timeline, and pricing. This will create a structured framework for the rest of the proposal.

Sample Prompt:

"You are a proposal writer outlining a proposal for furnishing a home with eco-friendly products. Provide a basic structure for the proposal, including sections such as Introduction, Project Scope, Objectives, Deliverables, Timeline, Pricing, and Next Steps."

Save the proposal outline in your AI Resource Center under Proposal Templates.

4. Draft the Project Scope and Objectives

Use AI to create a detailed description of the project scope and objectives. This section should outline the specific furnishings and decor the proposal will cover and the goals for enhancing the home's style and sustainability.

Sample Prompt:

"You are a proposal writer drafting the Project Scope and Objectives sections for an interior design proposal. The client is an interior designer furnishing a high-end residential property. Describe what our eco-friendly home goods will cover (e.g., living room, bedroom, kitchen) and outline specific objectives like enhancing aesthetics, promoting sustainability, and providing durability."

Save this draft in your AI Resource Center under Proposal Drafts.

5. Prompt AI to Gather Information for Deliverables, Pricing, and Timeline

Use AI to ask you questions that will help fill out the deliverables, pricing, and timeline sections. By answering these questions, you'll provide the AI with enough context to draft these sections accurately.

Sample Prompt:

"You are a proposal writer preparing to draft the Deliverables, Pricing, and Timeline sections of a proposal to furnish a home with eco-friendly products. Ask me questions to gather details about the specific products, quantities, budget, and deadlines for the project."

Questions the AI might ask:

- "What specific rooms or areas will be furnished (e.g., living room, bedroom, kitchen)?"
- "Are there any key items or products the client wants included (e.g., sustainable sofas, reclaimed wood tables)?"
- "What is the client's budget range for this project?"
- "What timeline or deadlines are required for completing each room or area?"

Record your answers and save them in your AI Resource Center under Proposal Information so you can refer back to them.

6. Draft the Deliverables, Pricing, and Timeline Sections

Now that you've provided the AI with information about the project, use it to draft the Deliverables, Pricing, and Timeline sections based on your answers from the previous step.

Sample Prompt:

"You are a proposal writer drafting the Deliverables, Pricing, and Timeline sections for an interior design proposal to furnish a home with eco-friendly products. Based on the information provided already, write a detailed Deliverables section listing specific items, a Pricing section with a breakdown of costs, and a Timeline section outlining delivery and setup milestones."

Save the draft in your AI Resource Center under Proposal Drafts.

7. Draft a Compelling Conclusion and Next Steps

Use AI to write a strong closing section for your proposal, including a CTA. The conclusion should reinforce your enthusiasm about working with the designer and provide clear next steps for moving forward.

Sample Prompt:

"You are a client relations manager writing the conclusion and next steps for a proposal to an interior designer. Wrap up the proposal with an enthusiastic CTA, and include instructions for the client on how to proceed if they'd like to move forward with the furnishing project."

Refine and save the conclusion and next steps in your AI Resource Center under Proposal Templates.

Common Mistakes to Avoid

1. Not Customizing the Proposal to the Client's Specific Needs

Mistake: Using a generic proposal template without adjusting it for the client's unique goals and challenges, which can make the proposal feel impersonal.

Better: Tailor each proposal to highlight how your solution addresses the client's specific needs. Use AI to generate personalized content based on client insights or past project information, ensuring the proposal resonates with the client's vision.

2. Overloading with Technical Jargon or Industry Terms

Mistake: Including too much technical language, which can confuse or alienate clients who are unfamiliar with the industry's terms.

Better: Simplify the language in your proposal and focus on clear, benefit-driven explanations. AI can assist by rephrasing technical descriptions into client-friendly language, helping keep the proposal accessible and engaging.

3. Not Including a Clear Project Scope and Timeline

Mistake: Failing to outline the project's scope, milestones, or timeline, which can lead to misunderstandings about deliverables and deadlines.

Better: Provide a clear, structured project plan, including key milestones and estimated timelines. Ask AI to help organize the proposal into sections, ensuring each deliverable and its timeline are explicitly stated.

4. Using Generic Benefits Instead of Client-Specific Value

Mistake: Listing generic benefits that could apply to any client, which doesn't make the proposal feel tailored.

Better: Highlight specific value points based on the client's industry, goals, or challenges. AI can suggest benefits that resonate directly with the client's sector, making the proposal more relevant and compelling.

5. Omitting Visual Elements to Enhance Readability

Mistake: Submitting a text-heavy proposal without visuals, which can reduce readability and engagement.

Better: Include charts, graphs, or images to visually support key points. You can prompt AI to suggest where visuals might be useful, or even request ideas for data visualization to make the proposal more impactful.

6. Failing to Include Next Steps and Contact Information

Mistake: Leaving out a clear CTA or contact details, which can make it harder for the client to proceed.

Better: Conclude with a strong CTA and provide direct contact information. Use AI to suggest next-step language that encourages prompt client action, helping you move quickly from proposal to project.

Upload your draft proposal or a screenshot of its main sections, and ask AI to assess clarity, relevance, and tone. Request suggestions for enhancing readability, tailoring benefits, or adding visual elements. Use AI's feedback to refine your proposal into a clear, client-focused document that stands out.

Reflection for Day 23

How comfortable do you feel using AI to create a proposal for a specific client need?

What parts of the proposal writing process felt easiest with AI support?

How could you use these proposal templates for future projects or clients?

Day 24

Writing a High-Value White Paper or Guide

Goal:

Today, you'll use AI to outline and draft a white paper on the benefits of using eco-friendly furnishings for office buildings. This resource is designed to inform contractors and corporations about the advantages of sustainable office furnishings, helping to establish your brand as an authority in the eco-friendly space. By the end of today, you'll have a well-organized draft that can be used as a lead magnet or educational tool to engage potential business clients.

Activities for Day 24

1. Start a New Thread and Name It

Open a new thread in ChatGPT and name it **"Eco-Friendly Office Furnishings White Paper".** This will help keep prompts and outputs organized as you work through today's activities.

Reminder: If you haven't set up a Custom GPT yet, start each new thread by copying your Training Document, Brand Voice & Tone Guidelines, and Brand Persona from your AI Resource Center into the prompt box. This will ensure the AI aligns with your brand's tone, style, and audience throughout today's activities.

2. Define the Audience and Purpose

Use AI to clarify the target audience and purpose of the white paper. This will help you fine-tune the tone and focus the content on the specific needs of contractors and corporations.

Sample Prompt:

"You are a content strategist drafting a white paper for an eco-friendly home goods brand. The white paper's target audience is contractors and corporations interested in sustainable office furnishings. Write a summary of the audience's main concerns, such as cost savings, employee well-being, and environmental impact, to ensure our white paper addresses these points."

Save this audience profile in your AI Resource Center for future reference.

3. Brainstorm Key Benefits of Eco-Friendly Office Furnishings

Ask AI to list the main benefits of using eco-friendly furnishings in office buildings. These benefits could include improved employee health, lower environmental impact, cost savings through energy efficiency, and enhanced corporate reputation.

Sample Prompt:

"You are an environmental consultant brainstorming key benefits of eco-friendly office furnishings for a white paper. List benefits relevant to contractors and corporations, such as improved indoor air quality, reduced carbon footprint, long-term cost savings, and alignment with corporate sustainability goals."

Save these benefits in your **AI Resource Center**.

4. Outline the Structure of the White Paper

Use AI to generate a detailed outline for the white paper. The structure might include sections such as Introduction, Benefits Overview, Supporting Data, Case Studies, and Conclusion/Call to Action.

Sample Prompt:

"You are a content strategist outlining a white paper titled 'The Benefits of Eco-Friendly Furnishings for Office Buildings.' Provide an outline with sections for an Introduction, Overview of Key Benefits, Supporting Data and Research, Case Studies of companies that made the switch, and a Conclusion with a call to action."

Save the refined outline for future reference.

5. Draft the Introduction and Benefits Overview

Use AI to draft an engaging introduction that hooks the reader and briefly explains why eco-friendly furnishings are important for office spaces. Then, write the Benefits Overview section, which summarizes the key advantages for contractors and corporations.

Sample Prompt:

"You are a white paper writer drafting the Introduction and Benefits Overview for a guide on eco-friendly office furnishings. Write an engaging introduction that highlights the growing importance of sustainability in corporate spaces, followed by a section that summarizes the key benefits for contractors and corporations."

Save the introduction and benefits overview wherever you like to draft documents.

6. Gather Supporting Data, Case Studies, and Examples with Source Citations

Ask AI to provide relevant statistics, studies, case studies, and links to credible sources that support the benefits outlined in the white paper. Including data and citations will enhance credibility and make the content more compelling.

Sample Prompt:

"You are a research assistant gathering data for a white paper on eco-friendly office furnishings. Provide recent statistics on the impact of sustainable materials on indoor air quality, employee productivity, and long-term cost savings. Include any case studies of companies that successfully implemented eco-friendly furnishings in their offices. For each data point, include a citation with a link to the source."

Save the data, case studies, and citations in your white paper document folder.

7. Draft the Conclusion and Call to Action

Use AI to write a strong conclusion that reinforces the benefits of eco-friendly office furnishings and encourages the reader to take action. The call to action might invite contractors or corporations to consult with your company about eco-friendly options.

Sample Prompt:

"You are a content strategist drafting the conclusion and CTA for a white paper on eco-friendly office furnishings. Write a summary of the main points and provide a CTA encouraging contractors and corporations to contact us for a consultation or quote on sustainable furnishings."

Save the conclusion and call to action to your white paper draft.

8. Compile the Complete White Paper

Now that you have all the sections drafted, ask AI to combine them into a cohesive white paper. This will include the introduction, benefits overview, supporting data and case studies (with citations), and conclusion. Review the draft to ensure a smooth flow and consistent tone.

Sample Prompt:

"You are a white paper editor. Please compile the following sections into a single, cohesive white paper on the benefits of eco-friendly furnishings for office buildings. Include the introduction, benefits overview, supporting data with citations, case studies, and conclusion with a call to action. Ensure a consistent tone and smooth transitions between sections."

Upload a draft or outline of your white paper, a screenshot, or a link to the document for AI feedback. Request an assessment of clarity, data support, and practicality of takeaways. Use AI's suggestions to improve structure, language, or the effectiveness of the CTA to ensure your white paper leaves a strong, lasting impression.

Make any refinements and save the complete white paper draft for final review and editing.

9. Brainstorm Distribution Strategies for Lead Generation

Ask AI to help you brainstorm ways to distribute the white paper effectively with the goal of generating leads from contractors, interior designers, and corporations. Distribution strategies could include digital channels, partnerships, and targeted outreach.

Sample Prompt:

"You are a marketing strategist brainstorming distribution strategies for a white paper on eco-friendly office furnishings. Suggest ways to distribute this white paper to generate leads among contractors, interior designers, and corporations. Include ideas like email campaigns, partnerships with industry organizations, social media ads, and direct outreach."

Save these distribution strategies in your white paper document folder.

Common Mistakes to Avoid

1. Not Clearly Defining the Target Audience and Purpose

Mistake: Writing a white paper without a clear understanding of the audience and goals, which can lead to unfocused content.

Better: Define your target audience and the white paper's purpose upfront, such as providing insights for contractors or corporate clients interested in sustainable practices. AI can help clarify audience-specific concerns, ensuring that your white paper resonates with readers' needs.

2. Using Overly Complex or Technical Language

Mistake: Including complex terminology without explanations, which can confuse readers unfamiliar with the topic.

Better: Keep the language accessible, focusing on the core benefits and implications of eco-friendly practices. AI can simplify technical content or add definitions for industry terms, making the white paper clear and approachable.

3. Skipping Data and Credible Sources

Mistake: Writing without data to back up claims, which reduces the white paper's credibility.

Better: Include relevant statistics, research, and case studies to support key points. AI can help locate credible sources, suggest key metrics, or generate visual representations like charts to strengthen your arguments.

4. Failing to Organize Content for Readability

Mistake: Presenting information in dense paragraphs without section breaks, which can make content overwhelming.

Better: Use headings, bullet points, and visuals to break up text and highlight key points. AI can assist by structuring the outline and formatting for readability, ensuring each section flows logically.

5. Omitting Practical Takeaways

Mistake: Providing general information without actionable insights, which may leave readers unclear about next steps.

Better: Include a summary or list of actionable takeaways relevant to the audience, such as "Implement energy-efficient lighting" or "Use sustainably sourced materials." AI can suggest practical tips based on each section, making the white paper more useful.

6. Missing a Strong Call to Action

Mistake: Ending the white paper without guiding readers on how to proceed, such as contacting your company for a consultation.

Better: Include a clear CTA that aligns with the white paper's goal, whether it's to explore products, book a consultation, or read related resources. AI can suggest CTA language that encourages follow-up and engagement with your brand.

Reflection for Day 24

What new insights did you gain about your target audience (contractors and corporations) through this exercise?

How comfortable do you feel using AI to develop a structured and persuasive white paper?

What other topics might work well for a white paper or guide aimed at corporate clients?

DAY 25

Product Launch Campaign Planning

Goal:

Today, you'll use AI to create a comprehensive product launch campaign for a new product. In this example, we'll be working on a launch plan for a new line of eco-friendly kitchenware, featuring pots and pans that take up 50% less space, are non-toxic, nonstick, and ethically manufactured. If you have a different product you're preparing to launch, feel free to substitute it in and follow the same steps to create your campaign.

Your campaign will include teaser content, launch-day messaging, follow-up engagement, optimized product descriptions, and a webinar outline to support the launch event. By the end of today, you'll have a well-rounded launch plan that maximizes visibility and excitement for your new product.

Activities for Day 25

1. Start a New Thread and Name It

Open a new thread in ChatGPT and name it **"Eco-Friendly Kitchenware Product Launch Campaign".** This will help keep prompts and outputs organized as you work through today's activities.

Reminder: If you haven't set up a Custom GPT yet, start each new thread by copying your Training Document, Brand Voice & Tone Guidelines, and Brand Persona from your AI Resource Center into the prompt box. This will ensure the AI aligns with your brand's tone, style, and audience throughout today's activities.

2. Describe the Product and Gather AI's Clarifying Questions

Begin by providing the AI with a brief description of the product to ensure it understands the unique features. Then, ask the AI to respond with any clarifying questions it may need to better support the campaign planning process.

Sample Prompt:

"Our new product is an eco-friendly kitchenware line that includes pots and pans. Key features include a space-saving design that reduces storage needs by 50%, non-toxic and non-stick materials, and ethical manufacturing practices. What clarifying questions do you have to help you better understand the product and assist with the campaign planning?"

Answer the AI's clarifying questions and save the conversation in your AI Resource Center for reference.

3. Brainstorm Teaser Content Ideas

Use AI to brainstorm ideas for teaser content that will build anticipation in the weeks leading up to the launch. Teaser content might include sneak peeks, countdowns, and "coming soon" graphics that hint at the product's space-saving, non-toxic, and eco-friendly qualities.

Sample Prompt:

"You are a marketing strategist brainstorming teaser content for a new eco-friendly kitchenware line launch. Generate ideas for social media posts, email snippets, and visuals that build excitement and curiosity about the product's space-saving design and non-toxic materials without revealing the full product details."

Save the teaser content ideas in your AI Resource Center

4. Draft an Optimized Product Description

Ask AI to help you craft a product description that highlights the unique features and benefits of your new kitchenware line. Focus on keywords that will resonate with both customers and search engines to maximize reach.

Sample Prompt:

"You are a product copywriter creating an SEO-friendly product description for a new line of eco-friendly kitchenware. Highlight its space-saving design, non-toxic materials, nonstick surface, and ethical manufacturing process. Include keywords like 'eco-friendly kitchenware,' 'non-toxic cookware,' and 'space-saving pots and pans.'"

Save the product description in your AI Resource Center for future reference.

5. Create Launch-Day Messaging

Use AI to draft the messaging for launch day. This might include an announcement email, social media posts, and a short press release. The goal is to generate excitement and encourage people to take immediate action, like visiting your website or placing an order.

Sample Prompt:

"You are a content marketer crafting launch-day messaging for a new eco-friendly kitchenware line. Write a launch announcement email, a series of social media posts, and a short press release that emphasize the product's space-saving design, non-toxic materials, and ethical production."

Save the launch-day messaging in your **AI Resource Center** for future reference.

6. Plan Follow-Up Engagement

Ask AI to help you create a follow-up engagement plan to keep the momentum going after the initial launch. This could include additional social media posts, customer testimonials, influencer partnerships, or limited-time offers.

Sample Prompt:

"You are a marketing strategist planning follow-up engagement for a product launch. Develop ideas for keeping the conversation going after launch day, such as customer testimonial posts, influencer partnerships, or limited-time discounts for the eco-friendly kitchenware line."

Save the follow-up engagement ideas you want to use in your AI Resource Center.

7. Draft a Webinar Outline to Support the Launch

If you're planning a webinar to introduce the product in more detail, use AI to draft an outline for the event. The webinar might include sections like product features, benefits, a live demonstration, and a Q&A session.

Sample Prompt:

"You are a webinar host planning an event to support the launch of a new eco-friendly kitchenware line. Create an outline that includes an introduction, product features (space-saving, non-toxic, nonstick), a live demonstration, and a Q&A session."

Save the webinar outline in your AI Resource Center.

8. Generate Targeted Social Media Ad Copy

Use AI to create targeted ad copy for social media platforms like Facebook, Instagram, and LinkedIn. These ads should be designed to reach specific audiences who are likely to be interested in eco-friendly home goods, such as environmentally conscious consumers and professionals in sustainable industries.

Sample Prompt:

"You are a social media ad copywriter creating targeted ad copy for a new eco-friendly kitchenware line launch. Write variations for Facebook, Instagram, and LinkedIn ads that highlight the product's space-saving design, non-toxic materials, and ethical manufacturing to appeal to eco-conscious consumers and businesses."

Save the social media ad copy to use when you launch the campaign.

9. Brainstorm Influencer and Partnership Opportunities

Ask AI to help you identify potential influencers, bloggers, or organizations in the eco-friendly or kitchenware space who could promote your new product. Collaborating with the right partners can significantly extend your product's reach.

Sample Prompt:

"You are a marketing strategist brainstorming potential influencers and partnership opportunities for an eco-friendly kitchenware line launch. Suggest influencers, bloggers, or organizations that focus on sustainability, eco-friendly living, or kitchen design who might be interested in promoting this product."

Save the influencer and partnership ideas in your **AI Resource Center for future reference.**

Common Mistakes to Avoid

1. Not Aligning Teasers and Messaging with the Product's Key Selling Points

Mistake: Launch content that doesn't emphasize the unique features or benefits, which can weaken initial engagement and interest.

Better: Center your teasers and messaging around what sets your product apart, such as its eco-friendly, space-saving design. AI can help brainstorm phrasing that spotlights these unique attributes, ensuring that early messages convey strong, memorable benefits.

2. Using Overly Revealing Teaser Content

Mistake: Teasers that give away too many product details, leaving little room to build anticipation.

Better: Plan teasers that hint at benefits without fully disclosing them. Ask AI to create subtle, curiosity-driven content like close-up images or countdown posts, which engage the audience's interest and anticipation ahead of the full reveal.

3. Creating Generic Product Descriptions

Mistake: Writing a standard product description that misses the emotional or practical impact on the customer.

Better: Craft a description that connects the product's features to specific customer needs. AI can assist by generating language that links features (e.g., "non-toxic, nonstick") to the audience's desire for healthier, space-efficient cookware.

4. Overlooking SEO in Product Descriptions

Mistake: Neglecting to add SEO keywords to the product description, which can limit search visibility.

Better: Use keywords related to the product's unique qualities, such as "eco-friendly kitchenware" or "space-saving pots and pans." AI can suggest keywords based on current search trends, increasing the product's discoverability online.

5. Forgetting a Pre-Launch Engagement Plan

Mistake: Skipping a plan to engage potential customers before launch, resulting in missed opportunities to build momentum.

Better: Create a pre-launch strategy that includes interactive content, such as polls or sneak peeks, to keep the audience engaged. AI can help schedule content or draft interactive questions, building excitement as launch day approaches.

6. Lack of a Clear Post-Launch Follow-Up Strategy

Mistake: Ending the campaign after launch without follow-up, which can reduce the product's long-term visibility and sales.

Better: Plan post-launch follow-up activities, like thank-you emails, user-generated content prompts, or product tutorials. AI can help generate follow-up messages or map out post-launch content ideas that maintain customer interest over time.

Upload your product launch plan, take a screenshot, or share a link to your campaign content. Ask AI to review the sequence, clarity, and impact of your teasers, product descriptions, and engagement strategies. Use AI's feedback to strengthen your pre-launch and post-launch efforts for maximum customer interest and retention.

Reflection for Day 25

How well did AI help you organize and plan a structured launch campaign?

What teaser content idea do you think will build the most excitement for this product?

How could you use similar launch strategies for other products in your line?

DAY 26

PR and Media Outreach, Including Guest Blogging & Podcasting

Goal:
Today, you'll use AI to develop a PR and media outreach plan focused on promoting your brand as a leader in eco-friendly home decor. Rather than pitching a single product, you'll be positioning your brand as a go-to source for sustainable living and decor trends. This approach opens up broader media opportunities and allows you to connect with outlets interested in topics like green home design, sustainable materials, and eco-conscious lifestyle choices. By the end of today, you'll have a media outreach plan that helps you connect with relevant publications, blogs, and podcasts to amplify your brand's visibility in the eco-friendly decor space.

Note: If you have a specific product line or focus area, feel free to adapt the prompts accordingly.

Activities for Day 26

1. Start a New Thread and Name It

Open a new thread in ChatGPT and name it **"PR and Media Outreach for Eco-Friendly Home Décor".** This will help keep prompts and outputs organized as you work through today's activities.

Reminder: If you haven't set up a Custom GPT yet, start each new thread by copying your Training Document, Brand Voice & Tone Guidelines, and Brand Persona from your AI Resource Center into the prompt box. This will ensure the AI aligns with your brand's tone, style, and audience throughout today's activities.

2. Identify Relevant Media Contacts and Outlets

Ask AI to help you identify potential media contacts, publications, and online outlets that focus on eco-friendly living, interior design, and sustainable home decor. Look for journalists, bloggers, and editors who cover topics related to sustainable living, eco-conscious decor, and green design.

Sample Prompt:

"You are a PR strategist identifying relevant media outlets and contacts for an eco-friendly home decor brand. Suggest publications, blogs, or journalists who cover topics like sustainable living, eco-friendly home design, interior decor trends, or green building practices."

Double-check the list of media contacts to make sure they're still active and save it for future reference.

3. Write a Press Release on Eco-Friendly Home Decor Trends

Use AI to draft a press release that establishes your brand as a leader in eco-friendly home decor trends. The press release should cover topics such as the increasing demand for sustainable home furnishings, popular eco-friendly decor choices, and your brand's role in providing ethically made, non-toxic options.

Sample Prompt:

"You are a PR writer drafting a press release about the growing trend of eco-friendly home decor. Write a press release that highlights popular sustainable decor choices, explains why more consumers are choosing eco-friendly home furnishings, and positions our brand as a leader in this space."

Refine and save the press release.

4. Draft a PR Pitch Email for Journalists

Ask AI to help you craft a concise and compelling email pitch that can be sent to journalists and media contacts. The pitch should introduce your brand, explain why eco-friendly home decor is a timely and important topic, and invite the journalist to explore this trend with your brand as a resource.

Sample Prompt:

"You are a PR specialist drafting an email pitch to journalists about eco-friendly home decor trends. Write a brief email that introduces our brand's commitment to sustainable home decor, explains the growing consumer interest in eco-friendly design, and invites the journalist to cover this trend."

Save the email pitch in your AI Resource Center for future reference

5. Develop Guest Blog Pitch Ideas on Eco-Friendly Decor Topics

Use AI to brainstorm guest blog ideas that showcase your expertise in eco-friendly home decor and appeal to readers of popular sustainability or home decor blogs. Each pitch should propose a topic relevant to the blog's audience and subtly tie in your brand's perspective on sustainable design.

Sample Prompt:

"You are a content strategist developing guest blog pitch ideas on eco-friendly home decor. Suggest three blog topics that would be relevant for sustainability or interior design blogs, such as 'How to Create a Green Living Room,' 'The Benefits of Non-Toxic Furniture in the Home,' or 'Eco-Friendly Decor Choices for a Modern Home.'"

Refine and save these guest blog pitch ideas in your AI Resource Center to access later.

6. Create Podcast Pitch Ideas on Sustainable Decor

Ask AI to generate ideas for pitching your brand as a guest on podcasts that focus on topics like sustainable living, interior design, or eco-friendly home choices. Each podcast pitch should briefly describe a topic you could discuss that would provide value to listeners while highlighting your brand's expertise.

Sample Prompt:

"You are a PR strategist creating podcast pitch ideas focused on eco-friendly home decor. Suggest a few podcast topics that would appeal to shows focused on sustainability, home design, or eco-conscious living, such as 'Eco-Friendly Decor Tips for Every Room,' 'How to Make Your Home Greener with Sustainable Decor Choices,' or 'Trends in Non-Toxic and Ethically Made Furniture.'"

Refine and save these podcast pitch ideas for later use.

7. Brainstorm Additional Topic Ideas for Guest Blogs and Podcasts

Ask AI to help you brainstorm additional topic ideas for both guest blogs and podcasts that align with your expertise in eco-friendly home decor. These topics should cover broader themes in sustainable living, green interior design, and the benefits of eco-friendly materials.

Sample Prompt:

"You are a content strategist brainstorming additional topic ideas for guest blogs and podcasts to promote our brand's expertise in eco-friendly home decor. Generate topic ideas that highlight the benefits of sustainable materials, non-toxic furnishings, and eco-conscious interior design."

Save the best topic ideas in your AI Resource Center.

8. Draft Follow-Up Emails for Media Contacts

Ask AI to help you draft a follow-up email template to send to media contacts who haven't responded to your initial pitch. The follow-up should be friendly and professional, gently reminding them of your pitch and offering additional information if they're interested.

Sample Prompt:

"You are a PR professional drafting a follow-up email for media contacts who haven't responded to an initial pitch. Write a brief, friendly email that reminds them of our brand's focus on eco-friendly home decor and offers additional information or interview opportunities if they're interested."

Save the final follow-up email template in your AI Resource Center.

Common Mistakes to Avoid

1. Targeting the Wrong Outlets for Media Outreach

Mistake: Focusing on media outlets that don't align with your audience or brand, which can result in low response rates and minimal impact.

Better: Identify outlets that share your brand's values or appeal to similar audiences. AI can help by analyzing media lists and suggesting outlets that match your target demographic, improving the chances of positive coverage.

2. Sending Generic or Impersonalized Pitches

Mistake: Using a one-size-fits-all pitch that doesn't address the specific interests of each journalist or publication, which can lead to quick dismissals.

Better: Personalize each pitch based on the journalist's recent work or the publication's focus. AI can draft tailored pitches by incorporating specific references or unique angles, making your outreach more relevant and engaging.

3. Overloading Pitches with Excessive Information

Mistake: Including too much detail in your initial pitch, which can overwhelm journalists and reduce readability.

Better: Keep pitches concise, highlighting only the most compelling aspects of your story. Use AI to refine pitches to be brief and impactful, focusing on elements that are likely to capture attention quickly.

4. Neglecting to Include Clear Contact and Follow-Up Information

Mistake: Sending a pitch without clear instructions on how journalists can reach you, which can delay responses.

Better: Conclude each pitch with direct contact details and an invitation to reach out for further information. AI can help format this section for clarity, ensuring your contact information is easy to find.

5. Ignoring the Power of Social Media Connections

Mistake: Relying only on email pitches and missing the opportunity to engage with journalists or influencers on social media.

Better: Follow and engage with media contacts on platforms like Twitter or LinkedIn. AI can suggest ways to initiate connections or provide prompts for thoughtful comments that build relationships before pitching.

6. Not Tracking and Analyzing Pitch Performance

Mistake: Sending pitches without tracking which ones receive responses, making it hard to refine future efforts.

Better: Use AI to help create a tracking document that logs pitch performance, response rates, and feedback. Analyzing this data over time can reveal patterns and help improve your pitch strategy.

Upload a draft of your media pitch or a list of target outlets, and ask AI to assess the relevance and impact of each. Use AI's suggestions to refine your pitch language, target specific journalists, or identify additional media contacts likely to cover your story.

Reflection for Day 26

What was the most useful media outreach strategy you developed today?

How comfortable do you feel using AI to draft PR and media outreach materials?

Which blogs or podcasts would be the highest priority for your outreach efforts, and why?

DAY 27

Advanced Content Repurposing

Goal:

Today, you'll learn how to use AI to repurpose a substantial piece of content (such as the eco-friendly home decor white paper you created on Day 24) into multiple formats. Repurposing content allows you to reach different audience segments across various platforms without having to create entirely new material from scratch. By the end of today, you'll have several versions of the same content—tailored for social media, blogs, emails, and more—helping you extend the reach and impact of your original work.

Note: If you don't have a white paper, guide, or larger content piece to repurpose, you can substitute any other long-form content, such as a detailed blog post, case study, or report.

Activities for Day 27

1. Start a New Thread and Name It

Open a new thread in ChatGPT and name it **"Content Repurposing for Eco-Friendly Home Décor".** This will help keep prompts and outputs organized as you work through today's activities.

Reminder: If you haven't set up a Custom GPT yet, start each new thread by copying your Training Document, Brand Voice & Tone Guidelines, and Brand Persona from your AI Resource Center into the prompt box. This will ensure the AI aligns with your brand's tone, style, and audience throughout today's activities.

2. Upload the White Paper for Context

Before you start repurposing, upload the eco-friendly home decor white paper (or your chosen long-form content) to give the AI full context. This will allow it to extract key points and details accurately.

Sample Prompt:

"You are a content strategist helping to repurpose this white paper on eco-friendly home decor into multiple formats. Here is the full text of the white paper: [paste white paper]."

After uploading, confirm that the AI has understood the content by asking it to summarize the main points. Save this summary in a project folder.

3. Summarize the White Paper into a Blog Post

Ask AI to summarize the key points of the white paper into a concise blog post. This blog post should retain the core message and highlights from the white paper but be accessible to a general audience.

Sample Prompt:

"You are a content marketer summarizing a white paper on eco-friendly home decor into a blog post. Write a 600-word blog post that covers the main points of the white paper, focusing on the benefits of sustainable materials, non-toxic furniture, and eco-conscious interior design."

Save the blog post in your project folder.

4. Create Social Media Snippets

Use AI to create several short, engaging snippets for social media platforms like Facebook, Twitter, Instagram, and LinkedIn. Each snippet should capture a key insight or benefit from the white paper to spark interest and drive traffic to the full content.

Sample Prompt:

"You are a social media strategist creating snippets based on a white paper on eco-friendly home decor. Write five short, attention-grabbing posts for social media that highlight different benefits of sustainable home decor, such as healthier indoor air quality, reduced environmental impact, and stylish, non-toxic furniture options."

Save the social media snippets in your project folder.

5. Draft an Email Newsletter Highlighting Key Insights

Ask AI to help you draft an email newsletter that highlights key insights from the white paper. The email should summarize the benefits of eco-friendly home decor and encourage subscribers to read the full white paper or blog post for more details.

Sample Prompt:

"You are an email marketer drafting a newsletter based on a white paper on eco-friendly home decor. Write a summary of the main points, and encourage readers to click through to the full content. Make the email engaging and informative, with a call to action to learn more about sustainable decor options."

Refine and save the email newsletter in your project folder.

6. Turn Key Points into an Infographic Outline

Use AI to outline an infographic based on the white paper's main points. The infographic should visually highlight the top benefits of eco-friendly home decor and could be used on your website, social media, or as a downloadable resource.

Sample Prompt:

"You are a graphic designer creating an infographic outline based on a white paper on eco-friendly home decor. Outline the key sections of the infographic, focusing on the top benefits of sustainable home decor, such as non-toxic materials, eco-friendly sourcing, and energy efficiency."

Save the infographic outline in your project folder.

7. Develop a Script for a Short Video or Reel

Ask AI to help you draft a script for a short video or Instagram Reel based on the white paper. The video could introduce eco-friendly home decor tips, highlight benefits, or provide a quick overview of sustainable design options.

Sample Prompt:

"You are a video content creator developing a script for a short video or Instagram Reel on eco-friendly home decor. Write a 30- to 60-second script that highlights a few quick tips for creating a sustainable and stylish home using eco-friendly furniture and decor."

Save the video script in your project folder.

8. Create a LinkedIn Article Based on the White Paper

Use AI to expand the white paper's core message into a LinkedIn article that positions your brand as a thought leader in sustainable decor. The article should be professional and informative, highlighting the importance of eco-friendly home decor in today's market.

Sample Prompt:

"You are a content marketer repurposing a white paper on eco-friendly home decor into a LinkedIn article. Write a professional article that highlights the value of sustainable home decor for health, the environment, and modern design trends. Position our brand as a leader in eco-friendly furnishings."

Save the final LinkedIn article in your project folder.

9. Draft Key Talking Points for a Webinar or Live Q&A

If you're planning a live Q&A or webinar about eco-friendly home decor, use AI to draft key talking points based on the white paper. These talking points will guide the discussion and ensure you cover the most important benefits and trends in sustainable home design.

Sample Prompt:

"You are a webinar host preparing key talking points for a live Q&A on eco-friendly home decor. Based on the white paper, outline points covering the benefits of sustainable materials, tips for choosing non-toxic furniture, and the latest trends in eco-friendly home design."

Save the webinar talking points in your project folder.

10. Repurpose Content into a Series of Pinterest Pins

Use AI to generate ideas for a series of Pinterest pins that visually showcase the benefits of eco-friendly home decor. Each pin could highlight a specific aspect of sustainable decor, such as non-toxic materials, energy-efficient designs, or eco-conscious living tips.

Sample Prompt:

"You are a Pinterest strategist creating ideas for a series of pins based on a white paper on eco-friendly home decor. Generate pin concepts that highlight different benefits of sustainable home decor, such as 'Why Non-Toxic Furniture Matters' and 'Top Tips for an Eco-Friendly Living Room.'"

Save the Pinterest pin ideas in your project folder.

Common Mistakes to Avoid

1. Trying to Repurpose Content Without Clarifying New Audience Needs

Mistake: Repurposing content without adjusting for a new audience or platform, which can make the message feel off-target.

Better: Define the intended audience for each new format, whether it's social media followers, newsletter subscribers, or blog readers. Use AI to identify the key points that resonate with each specific audience, tailoring content to suit their preferences.

2. Over-Reliance on Direct Copy-Pasting Without Adjusting Tone or Format

Mistake: Copying and pasting text without changing the tone or structure to fit the new format, which can reduce engagement.

Better: Adapt each repurposed piece to match the platform's tone and style. AI can assist by adjusting tone—for example, making LinkedIn posts more professional and Instagram captions more conversational.

3. Ignoring Platform Best Practices for Content Length and Structure

Mistake: Using the same content length or structure across all platforms, which can be less effective on certain channels.

Better: Keep in mind that platforms have different norms; short, punchy content works well on Twitter, while in-depth analysis is better suited for blogs. Use AI to help summarize or expand content based on the format's needs.

4. Missing Opportunities to Reinforce Key Visuals or Graphics

Mistake: Failing to incorporate visuals, which can make the repurposed content feel less engaging or less informative.

Better: Create visual elements, such as infographics, social media images, or slide summaries, that complement each format. AI can suggest ideas for visuals, ensuring they enhance the message and fit the platform's style.

5. Overlooking SEO Adjustments for Blog and Web Content

Mistake: Repurposing content for a blog or website without optimizing it for relevant SEO keywords, which can limit search engine visibility.

Better: Identify and incorporate keywords that align with the new format's focus. AI can suggest popular keywords based on current trends, boosting the discoverability of your repurposed content.

6. Not Setting Up Tracking to Measure Each Format's Effectiveness

Mistake: Repurposing content across formats without tracking performance, which can make it difficult to identify successful approaches.

Better: Set up performance metrics, such as engagement rates for social media and click-through rates for emails. Use AI to create a tracking document or to analyze performance data, allowing you to refine your repurposing strategy over time.

Upload the main content you're repurposing, or take a screenshot of your draft versions, and ask AI to review each piece's tone, length, and clarity for the intended platform. Request feedback on how to optimize visuals or keywords for SEO, and use AI's insights to refine each format for better reach and engagement.

Reflection for Day 27

Which content formats do you think will have the greatest impact for reaching your audience?

How comfortable do you feel using AI to repurpose content across different platforms?

What other pieces of content in your business could you repurpose using these same techniques?

DAY 28

Creating a Presentation

Goal:

Today, you'll use AI to help you create a visually engaging and informative presentation based on multiple sources, including the eco-friendly home decor white paper and any other relevant content. This presentation can be used for a webinar, client pitch, internal training, or any other setting where you need to showcase your expertise. By the end of today, you'll have a slide deck that effectively communicates your key points and aligns with your brand's tone and style.

Note: If you're not using the white paper, substitute other relevant documents, reports, or web content that you want to present.

Activities for Day 28

1. Start a New Thread and Name It

Open a new thread in ChatGPT and name it **"Eco-Friendly Home Decor Presentation Creation".** This will help keep prompts and outputs organized as you work through today's activities.

Reminder: If you haven't set up a Custom GPT yet, start each new thread by copying your Training Document, Brand Voice & Tone Guidelines, and Brand Persona from your AI Resource Center into the prompt box. This will ensure the AI aligns with your brand's tone, style, and audience throughout today's activities.

2. Upload Source Documents and Provide URLs for Context

Begin by uploading multiple documents, pasting key sections, or inputting URLs to provide the AI with full context for creating the presentation. This will allow it to pull information from various sources to build a well-rounded slide deck.

Sample Prompt:

"You are a presentation designer helping me create slides based on several sources. Here are the main documents and URLs for context: [upload longer documents and paste key sections from documents or input URLs]. Please review these to understand the main points for the presentation on eco-friendly home decor."

Confirm the AI's understanding by asking it to summarize the main points across all sources. Save this summary in a project folder.

3. Outline the Presentation Structure

Ask AI to help you create a slide-by-slide outline for the presentation. Typical sections might include an Introduction, Key Benefits, Supporting Data, Case Studies, and a Conclusion/Call to Action.

Sample Prompt:

"You are a presentation strategist outlining a slide deck based on these documents about eco-friendly home decor. Create a slide-by-slide outline, including sections for an Introduction, Benefits of Eco-Friendly Decor, Supporting Data, Practical Tips, and Conclusion with a call to action."

Save the outline in your project folder.

4. Write Slide Titles and Key Points

Use AI to generate slide titles and key bullet points for each slide based on the outline. These should be concise, informative, and easy to read at a glance.

Sample Prompt:

"You are a presentation designer drafting slide titles and key points for a presentation on eco-friendly home decor. Based on this outline [paste outline], write slide titles and 2-3 bullet points per slide."

Refine and save the slide titles and key points in your project folder.

5. Draft Speaker Notes and Full Presentation Script

Ask AI to create brief speaker notes that expand on each slide's content. These notes will guide you if you're presenting live. Additionally, ask AI to combine the speaker notes into a full script that can be read word-for-word for a polished, live presentation or recorded voiceover.

Sample Prompt:

"You are a presentation coach drafting speaker notes for a presentation on eco-friendly home decor. Write speaker notes for each slide based on the titles and bullet points provided. Then, combine the speaker notes into a full, cohesive script that can be used for a live presentation or recorded voiceover."

Save both the speaker notes and the full script in your project folder.

6. Generate Visual Ideas for Each Slide

Ask AI to suggest visual elements that would complement the content of each slide. These could include icons, stock images, graphs, or infographics. Visuals will help make the presentation more engaging and memorable.

Sample Prompt:

"You are a graphic designer creating visual ideas for a presentation on eco-friendly home decor. For each slide, suggest visuals such as icons, stock photos, or infographics that would support the content and engage the audience."

Save the visual ideas in your project folder.

7. Design an Eye-Catching Title Slide and Conclusion Slide

Use AI to help you brainstorm design ideas for a strong title slide and a compelling conclusion slide. The title slide should set the tone, while the conclusion slide should leave a lasting impression and include a call to action.

Sample Prompt:

"You are a presentation designer creating ideas for a title slide and conclusion slide for a presentation on eco-friendly home decor. Suggest a layout, imagery, and wording for each slide to make a strong first and last impression."

Save the title and conclusion slide ideas in your project folder.

8. Ask AI to Compile the Presentation Text into a Full Slide Deck

Now that you have all the slide titles, key points, speaker notes, and visual ideas, ask AI to compile this information into a slide-by-slide text format. You can then use this as a reference to build the slide deck in PowerPoint, Google Slides, or another presentation tool.

Sample Prompt:

"You are a presentation editor. Please compile the slide titles, bullet points, speaker notes, and visual suggestions into a full slide deck text outline for a presentation on eco-friendly home decor. Format it in a way that I can easily transfer into presentation software."

Review carefully, ask for any refinements, and save the compiled presentation text outline in your project folder.

9. Brainstorm Distribution Channels for the Presentation

Ask AI to suggest ways to share and distribute the presentation to reach a broader audience. Ideas might include hosting a webinar, sharing slides on LinkedIn, uploading to SlideShare, or incorporating it into a sales pitch.

Sample Prompt:

"You are a marketing strategist brainstorming distribution channels for a presentation on eco-friendly home decor. Suggest ways to share this presentation to maximize reach and generate interest, such as through webinars, LinkedIn posts, SlideShare, or as a downloadable lead magnet."

Save the distribution ideas in your folder.

Common Mistakes to Avoid

1. Not Starting with a Clear Outline for the Presentation

Mistake: Jumping into slide creation without a structured outline, which can lead to disjointed or repetitive content.

Better: Begin by mapping out a slide-by-slide outline. This helps keep the presentation focused and ensures a logical flow. AI can help generate an initial outline or suggest a typical structure based on your topic, keeping the presentation cohesive and organized.

2. Using Dense Text on Slides

Mistake: Filling slides with too much text, which can overwhelm the audience and reduce readability.

Better: Summarize key points into concise bullet points or short phrases. Use AI to condense lengthy paragraphs into easy-to-read slide text, enhancing clarity and engagement for viewers.

3. Ignoring Visual Appeal

Mistake: Creating slides that lack visual elements, making the presentation feel monotonous or hard to follow.

Better: Add visuals such as icons, images, or data visualizations to support key points. AI can suggest relevant graphics or even generate ideas for effective visual layouts, ensuring each slide is visually engaging.

4. Not Tailoring Content to the Audience's Knowledge Level

Mistake: Including overly complex or basic information that may not resonate with the audience, potentially causing disengagement.

Better: Adjust the depth of content based on the audience's expertise. AI can help customize language or suggest specific examples to better connect with the audience's level of understanding.

5. Missing Interactive Elements

Mistake: Presenting in a one-way format without engaging the audience, which can limit interest and retention.

Better: Incorporate interactive elements like polls, questions, or calls for input. AI can recommend ideas for interactive segments, making it easy to add audience participation points at strategic intervals.

6. Failing to Include a Strong Call to Action

Mistake: Ending the presentation without a clear CTA, which can leave the audience uncertain about the next steps.

Better: Conclude with a CTA that invites the audience to connect, learn more, or act on what they've learned. AI can suggest CTA language that is relevant and engaging, encouraging meaningful follow-up.

Upload your presentation outline or slides, or provide a link for AI to review. Request feedback on clarity, visual impact, and engagement level. Use AI's insights to refine slide text, add visuals, and strengthen your CTA, ensuring a polished, impactful final presentation.

Reflection for Day 28

How comfortable do you feel using AI to create a structured and engaging presentation?

Which distribution channel do you think will have the most impact for sharing this presentation?

What other content topics could you turn into presentations to engage your audience?

DAY 29

Creating a Lead Generation Funnel, Part 1: Strategy & Lead Magnet Creation

Goal:

Today, you'll use AI to strategize the initial stages of a lead generation funnel targeted at interior designers, with the goal of converting them into regular customers who use your products to furnish their clients' homes. The funnel will consist of ads and content leading to a landing page, where interior designers can download a valuable lead magnet in exchange for their contact information. By the end of today, you'll have a clear funnel strategy and a completed lead magnet that addresses designers' specific needs and interests.

Note: If you're focusing on a different product or service, feel free to substitute it in and follow the same steps to create your customized funnel.

Funnel Overview:

- **Top of Funnel**: Ads and social media content to attract interior designers.
- **Landing Page**: Offering a lead magnet that provides real value to designers.
- **Lead Magnet**: An educational or practical resource that speaks directly to designers' needs.
- **Nurture Sequence**: 5-email series (to be created on Day 30).
- **Webinar**: Educational session (to be set up on Day 30).
- **Conversion Goal**: Invite designers to schedule a video call with sales consultants.

Activities for Day 29

1. Start a New Thread and Name It

Open a new thread in ChatGPT and name it **"Lead Generation Funnel for Interior Designers—Strategy & Lead Magnet".** This will help keep prompts and outputs organized as you work through today's activities.

> **Reminder:** If you haven't set up a Custom GPT yet, start each new thread by copying your Training Document, Brand Voice & Tone Guidelines, and Brand Persona from your AI Resource Center into the prompt box. This will ensure the AI aligns with your brand's tone, style, and audience throughout today's activities.

2. Define the Target Audience and Pain Points

Ask AI to help you refine your understanding of interior designers as a target audience. Focus on their specific pain points, needs, and goals when it comes to sourcing eco-friendly home decor and furnishings for clients.

Sample Prompt:

"You are a marketing strategist defining the target audience for an eco-friendly home decor brand that wants to attract interior designers. Describe their main pain points and needs, such as finding sustainable products, managing budgets, meeting client demands for style, and needing reliable suppliers."

Save the audience profile in a new project folder.

3. Map Out the Funnel Stages and Key Messages

Use AI to help you map out each stage of the funnel, clarifying the purpose of each step and the key messages to convey to designers at each stage. This will guide the tone and content for the lead magnet, ads, landing page, and nurture sequence.

Sample Prompt:

"You are a funnel strategist outlining a lead generation funnel for an eco-friendly home decor brand targeting interior designers. Outline the purpose and key message for each stage: (1) Ads/Content, (2) Landing Page, (3) Lead Magnet, (4) Nurture Sequence, (5) Webinar, and (6) Sales Call CTA."

Save the funnel map in the project folder.

4. Brainstorm Lead Magnet Ideas for Interior Designers

Ask AI to suggest lead magnet ideas that would provide high value to interior designers. Examples might include a "Sustainable Sourcing Guide for Interior Designers," a "Budget-Friendly Eco-Decor Catalog," or a "Style Guide for Eco-Friendly Homes."

Sample Prompt:

"You are a lead generation strategist brainstorming lead magnet ideas to attract interior designers. Suggest lead magnets that would appeal to their needs, such as finding unique sustainable products, understanding eco-friendly certifications, or creating stylish, eco-conscious spaces for clients."

Save the lead magnet ideas to your project folder.

5. Select and Outline the Lead Magnet

Once you've selected a lead magnet idea, use AI to help you create a detailed outline for the content. The lead magnet should be both practical and visually appealing, providing actionable information that interior designers can use in their work.

Sample Prompt:

"We are creating a lead magnet titled '[Insert Lead Magnet Title Here].' Outline the main sections, such as an introduction to sustainable design, tips for finding eco-friendly suppliers, a list of sustainable certifications to look for, and product recommendations."

Save the lead magnet outline in your folder.

6. Draft the Content for the Lead Magnet

Ask AI to help you create the actual content for the lead magnet. Each section should be informative, engaging, and designed to position your brand as a valuable resource for interior designers who care about sustainability.

Sample Prompt:

"You are a content writer developing the lead magnet titled '[Insert Lead Magnet Title Here].' Write the content for each section based on this outline: [paste outline]. Make it informative and actionable, with tips and recommendations for sourcing eco-friendly home decor."

Upload your lead magnet content or provide a link to the draft for AI feedback. Request a review of clarity, relevance, and the strength of the CTA. Use AI's recommendations to refine the title, add keywords, and adjust the lead magnet format as needed for maximum impact.

Review carefully, ask for any refinements, and save the lead magnet content in a separate document in your project folder.

7. Brainstorm Visual Design Elements for the Lead Magnet

Ask AI to suggest design elements that will make the lead magnet visually appealing and aligned with your brand. This might include specific colors, font styles, icon suggestions, or layout ideas that make it look polished and professional.

Sample Prompt:

"You are a graphic designer creating ideas for the visual design of a lead magnet titled '[Insert Lead Magnet Title Here].' Suggest color schemes, font choices, icon ideas, and layout suggestions to make it visually appealing and on-brand."

Save the design ideas in your folder.

8. Draft the Landing Page Copy for the Lead Magnet

Now that you have the lead magnet content ready, ask AI to help you write the landing page copy. This page should briefly describe the lead magnet, highlight the value it provides to interior designers, and include a strong call to action inviting them to download it.

Sample Prompt:

"You are a landing page copywriter creating copy for a lead magnet titled '[Insert Lead Magnet Title Here].' Write persuasive copy that explains what the guide offers, how it helps interior designers, and encourages them to download it."

Save the landing page copy in your project folder.

9. Brainstorm Ad and Social Media Content Ideas to Promote the Lead Magnet

Ask AI to suggest ad and social media content ideas that will drive traffic to the landing page. The ads and posts should speak directly to interior designers, emphasizing the value of the lead magnet and piquing their interest.

Sample Prompt:

"You are a social media strategist creating ad and social media content ideas to promote a lead magnet for interior designers. Suggest ad headlines, social media post ideas, and targeting tips that would attract designers interested in sustainable sourcing."

Save the ad and social media content ideas in your project folder.

10. Select the Best Ad and Social Media Ideas, and Ask AI to Write the Final Posts

Review the list of ad and social media content ideas generated in Step #9. Choose the most effective ones, and ask AI to write the final versions. These should be engaging, targeted to interior designers, and optimized to drive traffic to the landing page.

Sample Prompt:

"Based on the ideas we brainstormed, write the final ad and social media posts to promote the lead magnet titled '[Insert Lead Magnet Title Here].' Make sure they are targeted to interior designers and include a strong call to action."

Review and ask for any refinements necessary, then save the final ads and social media posts in your project folder.

Common Mistakes to Avoid

1. Creating a Lead Magnet Without Clear Value to the Target Audience

Mistake: Developing a lead magnet that doesn't address specific challenges or needs, which can lead to low downloads and limited engagement.

Better: Focus on an actionable topic that solves a clear problem for your audience, such as "Sustainable Sourcing for Interior Designers." Use AI to generate ideas based on audience pain points, ensuring the lead magnet provides high value.

2. Using Overly Generic or Non-Compelling Titles

Mistake: Choosing a title that doesn't emphasize the unique benefits of the lead magnet, which can make it less attractive to potential leads.

Better: Create a title that highlights the value proposition directly, like "Eco-Friendly Design Guide: Top 10 Tips for Interior Designers." AI can help brainstorm title variations that emphasize appeal and clarity for maximum engagement.

3. Failing to Include a Strong Call to Action

Mistake: Leaving out a clear CTA, which can make it unclear what action the reader should take next, reducing conversion potential.

Better: Include a compelling CTA, such as "Download Now" or "Get Your Free Guide," to prompt immediate action. Use AI to create CTAs tailored to the funnel stage, making it easy for readers to understand the next step.

4. Neglecting Follow-Up Content for Lead Magnet Users

Mistake: Sending the lead magnet without planning a follow-up sequence, which can miss opportunities to deepen engagement.

Better: Develop a nurture email series that gradually introduces your brand's value and provides additional resources. AI can assist by drafting a sequence of personalized emails to build trust and encourage further action, like scheduling a consultation or joining a webinar.

5. Ignoring SEO and Keyword Optimization

Mistake: Creating the lead magnet and landing page without optimizing keywords, which can reduce search visibility.

Better: Identify SEO keywords relevant to your audience's interests, such as "sustainable interior design tips" or "eco-friendly decor." AI can help generate and incorporate keywords in the lead magnet and landing page copy to improve discoverability.

6. Not Testing Different Formats or Content Styles

Mistake: Using only one format (e.g., a PDF guide) without testing alternatives that might resonate more with the audience, such as video or checklist formats.

Better: Test different lead magnet styles, like video tutorials, downloadable checklists, or interactive quizzes, to see which performs best. AI can assist by suggesting format variations and helping adapt content to each.

Reflection for Day 29

How effective do you think today's lead magnet and landing page strategy will be in attracting interior designers?

What other types of resources or tools could you offer in future lead magnets to provide value to designers?

How comfortable do you feel using AI to create a strategy and build assets for a targeted lead generation funnel?

Day 30

Creating a Complete Lead Generation Funnel, Part 2: Nurture Sequence, Webinar, and CTA

IMPORTANT:

Today, you will continue working in the same thread from Day 29. This will keep all prompts, outputs, and context in one place, helping AI better understand the funnel and follow-up activities you're creating. Do not start a new thread.

Goal:

Today, you'll complete the lead generation funnel for attracting interior designers as regular customers. Building on the lead magnet and landing page from Day 29, you'll now use AI to create a 5-part email nurture sequence, set up an educational webinar, and design a compelling call to action that invites designers to schedule a video call with your sales team. By the end of today, you'll have a fully developed funnel ready to convert leads into loyal customers.

Reminder: At each step, feel free to ask AI for tweaks or adjustments. This is your opportunity to make sure each element aligns with your vision before moving on to the next part.

Activities for Day 30

1. Brainstorm Webinar Topic and Title

Use AI to help brainstorm a webinar topic and title that will be compelling to interior designers interested in eco-friendly home decor. The title should be engaging and promise valuable insights, such as sourcing tips, design inspiration, or sustainability best practices. Once you've chosen a topic and title, you can use it consistently in the email CTAs and throughout the funnel.

Sample Prompt:

"I'm targeting interior designers interested in eco-friendly home decor. Brainstorm engaging webinar topics and titles that provide valuable insights, such as sustainable sourcing tips, eco-friendly design inspiration, or best practices for creating environmentally conscious interiors. Please suggest five options with a focus on expertise and value for this audience."

Once you have a title you're happy with, ask AI for any refinements to improve its clarity or appeal. Save the final webinar title in your project folder for future reference.

2. Plan the 5-Part Email Nurture Sequence with Dual CTA

Ask AI to outline a 5-part email nurture sequence that will take designers from downloading the lead magnet to engaging with your brand and, ultimately, signing up for the webinar. Each email should have a specific purpose, such as introducing your brand, sharing valuable content, or addressing common designer challenges. Each email will also include a dual CTA: one to register for the webinar and another to schedule a direct call with a specialist if they're ready to speak sooner.

Sample Prompt:

"Outline a 5-part email nurture sequence for a lead generation funnel targeting interior designers interested in eco-friendly home decor. Each email should have a specific goal, such as introducing the brand, providing valuable tips, addressing challenges, or promoting the webinar. Include a dual CTA in each email: one to register for the webinar and one to schedule a call with a sales specialist."

Review the sequence to ensure it aligns with your funnel goals, and ask AI for refinements if necessary. Save the final email sequence outline in your project folder.

3. Draft Each Email in the Nurture Sequence with Dual CTA

Use AI to help you write each email in the nurture sequence. Each email should be engaging, build trust, and include a dual CTA that encourages designers to either register for the webinar or schedule a call directly with your team.

Sample Prompt:

"Write the first email in a 5-part nurture sequence targeting interior designers interested in eco-friendly home decor. The email should welcome them after downloading the lead magnet, introduce the brand, and include a dual CTA: one to register for a webinar on sustainable design tips and another to schedule a call with a sales specialist."

Once an email is finalized, repeat the prompt for the next one. When all emails are drafted, review them and ask AI to refine tone, structure, or wording as needed. Save the final versions of all five emails in your project folder.

4. Create the Webinar Outline

Now that your emails are driving interest in the webinar, use AI to help create a detailed outline for the webinar content. The outline should cover key sections, such as an introduction, main topics, a product showcase (if applicable), and a Q&A session. The webinar should deliver value by addressing interior designers' specific needs and challenges related to eco-friendly home decor.

Sample Prompt:

"Create a detailed outline for a webinar targeting interior designers interested in eco-friendly home decor. The webinar should include an introduction, 3-4 key sections on sustainable sourcing, design inspiration, and best practices, a short product showcase, and a Q&A session. Focus on delivering value and actionable tips."

Review the outline and ask AI for feedback or adjustments to ensure it flows well and addresses your audience's needs. Save the final webinar outline in your project folder.

5. Break Down the Webinar Outline into Individual Slides

With the webinar outline ready, ask AI to break down each section into individual slides. Each slide should cover a single key point, making it easy to follow along during the presentation.

Sample Prompt:

"Take the previous webinar outline and break it down into individual slides for a presentation. Each slide should focus on one main idea, with suggestions for slide titles and brief content for each section. Make sure the presentation flows logically."

Review the slides for clarity and ask AI to make adjustments where necessary. Save the final slide breakdown in your project folder.

6. Draft the Webinar Script or Speaker Notes

Ask AI to help you create a script or speaker notes for each slide. This script should be conversational and engaging, positioning your brand as a trusted resource for sustainable furnishings and eco-friendly decor.

Sample Prompt:

"Write a conversational webinar script or speaker notes for the slides. Each slide covers a key point related to eco-friendly home decor for interior designers. The tone should be professional but engaging, with clear transitions between topics."

Once the script or speaker notes are ready, review for tone and engagement, and ask AI for revisions if needed. Save the final script or speaker notes in your project folder.

7. Draft a Thank-You Email for Webinar Attendees with Replay Link

After the webinar, you'll want to follow up with attendees. Ask AI to draft a thank-you email that expresses appreciation, summarizes key points from the webinar, and includes a link to watch the replay. This email should also remind them of the opportunity to schedule a call with your team if they have further questions or want personalized guidance.

Sample Prompt:

"Draft a thank-you email for the webinar attendees. The email should express appreciation, summarize the main takeaways, include a replay link, and remind them to schedule a call with our sales team for personalized guidance."

Review the email and ask for refinements as needed. Save the finalized thank-you email in your project folder.

8. Upload Your Work for AI Feedback

Ask AI to assess each part's clarity, engagement, and alignment with audience interests. Use AI's feedback to fine-tune topics, simplify presentation slides, or craft follow-up messages that drive conversion.

Sample Prompt:

"Assess this webinar outline for clarity, engagement, and relevance to interior designers interested in eco-friendly home decor. Provide suggestions for improvement or refinement where necessary."

Incorporate AI's feedback into your materials, then save the final, refined versions in your project folder.

9. Generate a Comprehensive Implementation Checklist

As the final activity, ask AI to create a comprehensive, step-by-step implementation checklist that covers every part of the funnel. This checklist will help ensure that every element is in place before launching. Each item should have a checkbox, and at the end of each section, include a line for a supervisor to sign off on completion.

Sample Prompt:

"You are a project manager creating an implementation checklist for a lead generation funnel targeting interior designers. Create a sequential checklist that includes all necessary tasks from start to finish, with checkboxes for each item and a signature line for a supervisor to approve each section."

Review the checklist to ensure it's complete and clear, then save it in your project folder as the final document for this funnel.

Common Mistakes to Avoid

1. Choosing a Webinar Topic That Doesn't Align with Audience Needs

Mistake: Selecting a topic that doesn't directly address the specific interests or challenges of your target audience, which can lead to low attendance and engagement.

Better: Choose a topic that aligns with the pain points and goals of your audience. AI can assist by generating topic ideas based on popular trends, audience feedback, or frequently asked questions, ensuring that your webinar is relevant and compelling.

2. Using Generic Language in the Webinar Promotion

Mistake: Writing webinar promotion copy that sounds generic or uninspired, which can fail to capture attention.

Better: Craft copy that emphasizes the unique value attendees will gain from the session. AI can help brainstorm engaging headlines and descriptions that highlight key benefits and speak directly to audience needs.

3. Not Structuring the Nurture Sequence for Progressive Engagement

Mistake: Creating a nurture sequence that doesn't gradually build interest, potentially missing opportunities to deepen engagement.

Better: Design each email in the sequence with a specific purpose, leading recipients from initial awareness to active participation. Use AI to plan and draft emails that increase in detail and CTA strength over time, guiding recipients through the funnel effectively.

4. Overloading Webinar Content Instead of Focusing on Key Points

Mistake: Trying to cover too much in a single webinar, which can overwhelm attendees and reduce clarity.

Better: Focus on a few core points or tips that deliver high value. AI can assist by summarizing complex information into concise points and even suggesting slide-by-slide content that keeps the presentation streamlined and impactful.

5. Failing to Include an Engaging Q&A or Interactive Segment

Mistake: Presenting a one-way lecture without interactive elements, which can result in disengagement.

Better: Include time for a Q&A session or add interactive polls to maintain engagement and allow attendees to clarify questions. AI can suggest questions based on audience interests or provide poll ideas, making it easier to keep participants involved.

6. Neglecting Follow-Up Actions for Attendees

Mistake: Ending the webinar without a clear follow-up, which can reduce the potential for conversion.

Better: Send a thank-you email with a recording link and an invitation for a one-on-one call. AI can help draft this email, providing a strong call to action that encourages further interaction with your brand.

Reflection for Day 30

How comfortable do you feel now that you've built a complete lead generation funnel from start to finish?

What did you find most valuable about using AI to assist with this project?

Are there other types of audiences or funnels you'd like to build using a similar process?

WEEK 4 RECAP

PERSONALIZATION, AUTOMATION, AND FINE-TUNING YOUR AI WORKFLOW

Congratulations on completing Week 4 of your AI journey! This week, you explored how to use AI to personalize content at scale, automate repetitive tasks, and fine-tune your processes for maximum efficiency. These skills are essential for building systems that save you time while delivering meaningful, customized experiences to your audience. Let's review what you accomplished.

Key Skills and Accomplishments

1. **Advanced Personalization Techniques**
   - Learned how to use AI to craft highly personalized email content tailored to specific audience segments.
   - Practiced generating dynamic content variations, such as email subject lines, personalized openings, and customized calls-to-action.
   - Experimented with hyper-targeted messaging to engage individuals based on their unique preferences, behavior, or industry.
2. **Audience Segmentation and Targeting**
   - Used AI to help segment your audience by analyzing characteristics such as demographics, interests, or past interactions.
   - Generated messaging frameworks for each segment to ensure your communication speaks directly to their needs and challenges.

- Explored how to balance automation and personalization to scale your efforts without sacrificing quality.

3. **Content Automation and Workflow Optimization**
    - Automated repetitive content creation tasks, such as drafting routine email replies, follow-ups, and social media captions.
    - Streamlined workflows by integrating AI into your everyday processes, saving significant time and effort.
    - Created reusable AI prompts and templates to ensure consistency across projects.
4. **Fine-Tuning AI Output for Accuracy and Quality**
    - Practiced assessing AI-generated content for tone, clarity, and alignment with your brand voice.
    - Gave AI feedback on its outputs, learning how to refine prompts to achieve better results.
    - Developed a process for reviewing, editing, and finalizing content to maintain professional standards.
5. **Managing Large-Scale Campaigns with AI**
    - Learned to use AI as a partner in managing complex campaigns, including ideation, content creation, and follow-ups.
    - Practiced organizing projects by creating centralized systems to track AI outputs and save reusable resources.
    - Built a system to track and measure the performance of AI-generated content, ensuring it aligns with your business goals.

Reflection on Progress

Take a moment to reflect on how far you've come:

- How has AI changed the way you approach personalization and scaling your communication?
- Which automation tools or techniques saved you the most time this week?
- What new opportunities do you see for using AI to further enhance your marketing efforts?

Reflecting on these questions can help solidify the skills you've developed and give you clarity about how to move forward with confidence.

Common Challenges and Solutions

This week, you may have faced challenges like creating truly personalized messaging at scale or figuring out how to balance automation with maintaining a human touch. These challenges are common but solvable:

- When crafting personalized content, providing AI with specific details about audience segments (e.g., tone preferences, pain points) helps generate more relevant outputs.
- For automation, starting with small, repetitive tasks can help you build confidence and refine your systems before tackling larger projects.
- As always, iterative workflows—where you generate, review, and refine—are key to maintaining the quality of AI-generated content.

Looking Ahead

With Week 4 complete, you've mastered the core skills needed to personalize, automate, and refine your content creation processes. These capabilities are invaluable for scaling your efforts without sacrificing quality or impact.

Looking ahead, your next step is to implement what you've learned across your business or organization. Whether it's building detailed lead generation funnels, creating content at scale, or saving hours of time through automation, you now have the tools and knowledge to take full advantage of AI in your marketing and communication efforts.

Take a moment to celebrate your success! You've gone from beginner to someone capable of building efficient, high-impact AI workflows in just four weeks. That's a huge accomplishment, and it's only the beginning of what you can achieve. Keep pushing forward—you're on the path to becoming an AI-powered marketing pro!

CONCLUSION

CONGRATULATIONS ON COMPLETING YOUR AI JOURNEY

First of all, congratulations! Completing a 30-day journey is no small feat. You've dedicated time, focus, and energy to learning new skills, embracing AI, and building something tangible that will drive real value for your business or organization. That commitment to learning and growth is something to be incredibly proud of.

As you reflect on everything you've accomplished, know that this isn't just about completing tasks or following steps. Over the past month, you've taken significant strides in mastering AI as a tool, developing valuable skills, and pushing the boundaries of what's possible in your work. You didn't just learn about AI—you built a powerful, customized lead generation funnel, and equipped yourself with skills you can carry forward into all kinds of projects.

The Power of Prompt Engineering

Perhaps the most powerful skill you've gained in this journey is prompt engineering. The ability to clearly and strategically communicate your needs to an AI is a fundamental skill that will set you apart in an increasingly AI-driven world. You now know how to phrase what you want in a way that gets the best, most targeted results—no matter which platform you're using.

And that's the beauty of prompt engineering: it's platform-agnostic. Whether you're using ChatGPT, Claude, Copilot, Gemini, or another generative AI tool, you're now equipped with the skills to get high-quality outputs by crafting effective prompts. You've learned to break down tasks into logical steps, provide context, and refine AI outputs for precision—all of which make you a much more powerful user of this technology.

As AI continues to evolve, these skills will serve you well. New tools and models may emerge, but your ability to "speak AI" effectively will always give you an edge.

Recap of Key Skills and Accomplishments

In addition to prompt engineering, you've developed a toolkit of skills that will be invaluable in your business. Here's a quick recap of what you've learned:

- **Breaking Down Complex Projects**: You can now structure complex tasks—like building a full lead generation funnel—into manageable steps, and use AI to support each part of the process.
- **Brainstorming and Content Creation**: You've learned to prompt AI to generate ideas, write drafts, and refine content, making content creation faster and more engaging.
- **Marketing Strategy and Customer Journeys**: By mapping out customer journeys and creating targeted content, you've developed a deeper understanding of how to use AI to attract, nurture, and convert leads.
- **Testing and Refinement**: You know how to review AI outputs, make adjustments, and ask for tweaks, so that each piece of content aligns with your brand's voice and goals.
- **Implementation and Review**: You created a comprehensive, multi-stage funnel and ensured every element was in place and optimized before launch.

While you've used these skills to build a lead generation funnel in this book, they're **transferable to countless other projects**. Whether it's content repurposing, customer engagement, or even market research, you now have the foundational skills to bring AI into any area of your business.

Self-Assessment: Prompt Engineering Confidence Check

To help you reflect on your progress, here's a quick self-assessment. Rate your confidence level for each of the statements below, on a scale of 1 to 5, where one is "not confident" and five is "very confident."

Prompt Engineering Skills

- I feel confident in prompting AI to brainstorm content ideas across different topics. _______
- I know how to phrase a prompt to get specific, tailored outputs. _______
- I can effectively troubleshoot and refine AI outputs to improve quality. _______
- I feel comfortable guiding AI through multi-step tasks to produce polished content. _______

Customer Engagement and Retention Skills

- I can use AI to create personalized follow-ups, nurture sequences, and loyalty emails. _______
- I feel confident creating customer feedback forms, gathering insights, and analyzing responses. _______
- I understand how to use AI to structure client proposals and customer-facing case studies. _______
- I know how to apply customer feedback to improve products or services. _______

Strategic Content Creation and Brand Consistency Skills

- I can use AI to draft and optimize white papers, case studies, and other long-form content. _______
- I feel confident repurposing content across multiple channels and formats with AI. _______
- I know how to maintain brand voice consistency across social media, email, and ad copy. _______
- I can develop high-quality promotional content, including blogs, emails, and social media posts, using AI support. _______

Funnel Building and Marketing Strategy Skills

- I can use AI to create a multi-stage marketing funnel. _______
- I feel confident crafting nurture sequences and targeted emails using AI. _______
- I can build and optimize customer journeys with AI support. _______
- I understand how to develop a complete lead generation funnel, from top-of-funnel content to conversion-focused CTAs. _______

Market Research and Competitive Analysis Skills

- I can conduct market trends research and create insightful reports using AI. _______
- I know how to use AI to perform competitor research and identify industry trends. _______
- I feel confident in creating promotional strategies based on customer segments and market insights. _______

Reflecting on Your Growth

Consider your ratings for each category. Which skills do you feel most confident in, and which areas do you feel need more practice? Reflect on the steps you've taken and identify any further resources or exercises you may want to explore to continue building your AI-assisted marketing skills.

Reflection Notes:

Next Steps: Expanding Your AI Skills

As you look to the future, here are some ways to keep building on everything you've learned:

Experiment with Different AI Platforms: You now have a powerful set of prompt engineering skills, so try using them across other tools like Claude, Copilot, and Gemini. You'll notice slight differences in how each platform responds, which will deepen your understanding of generative AI's potential.

Apply AI to Other Business Areas: Consider ways to use AI outside of lead generation. For example, you could use it to generate social media content, answer customer inquiries, conduct market research, or brainstorm product development ideas.

Stay Adaptable and Keep Learning: AI is a rapidly evolving field, and new tools and models are always emerging. Keep testing, exploring, and adapting. The more you practice, the more comfortable you'll become as a leader in this space.

Inspiration and Vision for Future Success

As you continue experimenting with AI, think about the possibilities. Imagine how these tools can streamline your work, open up new creative avenues, and enable you to focus on strategic tasks. Here are a few ideas for future inspiring projects:

Content Refresh and Expansion: Use AI to audit your existing content, find opportunities for updates, and generate ideas for new pieces, creating a cohesive and evergreen content library.

Advanced Customer Segmentation: Prompt AI to help you create targeted customer segments, and then craft personalized email campaigns that speak directly to each group's needs.

Market Insights and Trend Analysis: Leverage AI to synthesize industry reports, summarize trends, and identify new opportunities for your business to stay ahead of the competition.

The potential is limitless—and now that you've built a strong foundation, you're equipped to use AI to its fullest.

Final Reflection: Celebrating Progress and Looking Forward

Take a moment to reflect on what you're most proud of from this journey. Maybe it's overcoming initial hesitation about AI, or maybe it's the feeling of accomplishment from creating a fully functional funnel. Write down one or two things you're most proud of:

What Am I Most Proud Of?

And what are you most excited to tackle next with your AI skills?

Embracing the Future of AI

As you move forward, know that you're now positioned as a leader in AI-powered marketing. You've gained skills that will make you adaptable, resourceful, and ready to embrace the future of technology. AI will continue to evolve, but you have the confidence to evolve right along with it.

Thank you for committing to this journey, for challenging yourself, and for building a foundation that will serve you well into the future. Congratulations again—you've accomplished something remarkable, and we can't wait to see where your AI skills take you next!

AI IS HERE—WHAT'S YOUR NEXT MOVE?

You've spent the last 30 days exploring how AI can transform the way you work. But that's just the beginning. Now it's time to practice, iterate, and keep learning.

- **Want to go deeper?**
  Our books break down AI adoption, strategy, and implementation—so you can stay ahead of the curve.
- **Need hands-on guidance?**
  From executive strategy workshops to AI implementation sessions, we help teams and leaders turn AI ideas into real business results.
- **Looking for more tools and resources?**
  Find practical guides, AI-powered templates, and expert insights to help you integrate AI effectively—whether for yourself, your team, or your entire company.

ABOUT THE AUTHOR

For over 20 years, Julie Anne Eason has been at the forefront of tech and digital marketing trends, helping companies bridge the gap between cutting-edge technology and business growth. With a deep understanding of the evolving tech landscape, she now leads MindFlare Consulting, where she assists businesses in leveraging AI to craft effective messaging, enhance branding, and scale efficiently.

Julie has worked with industry icons such as Russell Brunson and Alex Hormozi, giving her unique insight into the challenges companies face in marketing and lead generation. Her expertise goes beyond content creation; she transforms content into powerful, AI-driven strategies that generate leads, convert customers faster, and maximize profitability.

Under Julie's leadership, MindFlare Consulting empowers companies to harness AI, automate workflows, and boost ROI while maintaining authentic, on-brand messaging. MindFlare helps businesses embrace the future of technology, reduce overhead costs, and position themselves as leaders in a competitive global market.

To learn more, scan the code below to visit MindFlareConsulting.com or connect with Julie on LinkedIn at linkedin.com/in/julieanneeason.

ACKNOWLEDGMENTS

I would like to express my deepest gratitude to my mentors in the world of AI, Jeff J. Hunter and Melle Amade Melkumian. Thank you for showing me what's possible.

A special thank you to my publishing team, especially Kevin LeBlanc—I couldn't do what I do without you.

To my family, especially James and Ben, who never complain about cooking their own suppers when I'm on a deadline, thank you for your endless patience and support.

APPENDIX

ADVANCED TIPS FOR EFFECTIVE PROMPTING

Develop Your Ideal Expert Persona

Example: Rather than simply saying, "You are a content strategist," try, "You are an award-winning content strategist and expert copywriter known for crafting compelling, high-converting sales pages."

Picture your ideal expert advisor—whether that's a seasoned PR strategist, a visionary brand designer, or a social media influencer with proven success. Define their expertise clearly, so AI can "step into" this role and provide the nuanced advice you're looking for.

Be Specific in the Task Details

Example: Instead of, "Write a social media post," say, "Write a friendly, three-sentence Instagram post that highlights the eco-friendly benefits of our new product and includes a call to action to learn more in our bio."

AI performs best with concrete instructions. Specify the platform, tone, length, and goal of the content (e.g., engagement, awareness, or click-through). Think of the "What" stage as painting a picture—describe exactly what success looks like.

Add Context for Your Audience

Example: Instead of, "Explain the benefits of our product," say, "Explain the benefits of our eco-friendly cleaning products to health-conscious parents who care about non-toxic, natural household items."

Defining who the audience is (such as their pain points or values) allows AI to tailor the response, giving you more relevant and persuasive content.

Break Down Multi-Step Processes

Example: Rather than asking, "Write a complete blog post," try "Outline a blog post about the benefits of eco-friendly packaging. Include an introduction, three main points, and a conclusion. Then, draft the introduction paragraph based on the outline."

If a response is too general or lacks depth, break it into steps. You can then guide AI through each section with additional prompts, resulting in a more comprehensive final product.

Ask AI to Self-Check and Improve Outputs

Example: After receiving a response, say, "Review this email for tone and suggest improvements to make it more engaging and personable."

AI can critique and enhance its own work! Ask it to check for specific qualities like tone, readability, or clarity, then apply any suggestions to fine-tune your content.

Emphasize Style and Tone Preferences

Example: Instead of, "Write a promotional email," say, "Write a promotional email in a warm, conversational tone that feels personal and friendly, like it's coming from a helpful advisor."

Describe the tone, formality level, and style. Terms like "friendly," "authoritative," or "enthusiastic" guide the AI's language choices and keep content consistent with your brand's voice.

Experiment with Prompt Length and Phrasing

If the output isn't quite right, try rephrasing your prompt or adding specific phrases like "in three bullet points" or "in no more than 150 words." Sometimes, a small change in prompt structure will produce a much better result.

Request Supporting Data or Examples for Credibility

Rather than asking, "Write about the benefits of AI in business," prompt with, "Write a brief introduction to the benefits of AI in business, and cite at least three examples or recent statistics to support each point."

When the task requires credibility, prompt AI to incorporate statistics, case studies, or specific industry examples. AI can also recommend sources, but be sure to verify facts for accuracy.

Encourage Variations for Testing and Optimization

If you need variations, say, "Generate three different taglines for our eco-friendly brand, each with a unique emphasis: one on sustainability, one on quality, and one on affordability."

Testing multiple versions of content (like headlines, taglines, or emails) can reveal what resonates most with your audience. Asking for multiple versions in a single prompt saves time and inspires fresh ideas.

Adjust Tone or Style for Better Audience Fit

If a response sounds off, try, "Rewrite this blog introduction to be more casual and approachable for a general audience."

AI may default to a neutral or formal tone, so don't hesitate to request specific adjustments in tone, style, or even regional preferences (e.g., "Write in a tone that appeals to US-based professionals").

Re-Upload or Reference Key Documents for Consistency Across Threads

If you're switching threads and need consistency, re-upload any essential reference documents (like Brand Voice & Tone Guidelines or Training Documents) to keep AI responses on brand. Alternatively, create custom AI personas to avoid repetitive uploading.

Data Analysis and Research Tips

Define the Scope and Context of Your Data Analysis

Instead of, "Analyze this sales data," try, "Analyze this sales data to identify any trends in quarterly revenue growth and top-performing product categories."

Be specific about what you want to uncover, whether it's trends, anomalies, or comparisons. Setting a clear scope helps AI focus on the insights that are most relevant to your goals.

Provide a Summary of Key Data Columns or Variables

If you're working with data fields like "product category," "sales region," or "customer segment," list these fields in the prompt to guide the AI's analysis.

When uploading complex data sets, give AI a quick overview of the variables or columns, especially those relevant to your analysis. This makes it easier for AI to interpret and respond accurately.

Ask for Multiple Perspectives or Angles in Data Analysis

Instead of asking, "What insights can you provide from this data?" try, "Analyze this data set from three perspectives: customer demographics, purchasing behavior, and seasonal trends."

Approaching data from multiple angles often reveals more nuanced insights. Asking AI to consider different perspectives ensures that you cover various aspects of the data and gain a comprehensive view.

Request Actionable Insights Based on Findings

After an initial summary, prompt, "Based on these findings, suggest three actionable steps we can take to increase customer retention."

AI can help translate raw insights into practical actions. Ask it to identify specific steps or recommendations based on the analysis, which can guide decision-making.

Validate AI-Generated Statistics or Data Sources

If AI provides statistics or references, follow up by asking it to clarify the sources or add, "Please provide sources or data citations with links." Remember that while AI can suggest data, it may not have real-time sources, so double-check facts from reliable external references.

Ask AI to Explain Trends or Anomalies in Simple Terms

Rather than just accepting a chart or data point, prompt AI to "Explain the significance of the upward trend in Q3 sales and what it may indicate about customer demand."

Asking AI to explain trends or outliers helps you interpret results more intuitively. This is especially useful when presenting findings to stakeholders who may not have a technical background.

Use Hypothetical Scenarios for Predictive Analysis

"If we increase our marketing budget by 15%, how might that affect our Q4 sales?"

AI can model "what-if" scenarios that help predict outcomes of certain business decisions. Specifying hypothetical changes can help you explore different options and their potential impact.

Request Comparative Analysis for Benchmarks

Ask, "Compare our Q2 and Q3 sales to see if there's any seasonal pattern," or "How does this year's data compare to last year's?"

Comparative analysis is helpful for spotting patterns, assessing growth, and understanding performance over time. Comparing different data sets or time periods can give a clearer picture of trends or seasonal changes.

Generate a Research Summary with Key Takeaways

"Summarize the key findings from this research, highlighting the top three takeaways and any unexpected results."

After a detailed analysis, ask AI to condense findings into a high-level summary. This is especially helpful for quickly conveying insights to stakeholders or team members who need the core points without diving into the full data.

Include Context for Research Prompts to Improve Relevance

Instead of, "Provide market trends in eco-friendly products," say, "Identify the top three market trends in eco-friendly home goods for the past two years, focusing on consumer purchasing behaviors."

The more context you provide in research prompts, the more accurate and relevant the output will be. Including timeframes, geographic focus, or specific sectors can help AI deliver focused insights.

Prompt Refinement and Optimization Tips

Get AI Feedback to Improve Your Prompts

After writing your best prompt, ask AI to "act as a prompt engineer" and critique it. Try saying, "Review the following prompt and suggest ways to improve it. Write out the optimized version and explain why you made each change."

This method not only improves your current prompt but also helps you learn prompt optimization for future tasks by understanding the adjustments AI recommends.

Ask "What Else?" to Get Additional Insights

After a response, say, "What else can you think of?" or "What other perspectives can we explore?"

AI can often provide additional ideas or alternative angles that may not come to mind right away. Asking "What else?" encourages a more comprehensive exploration of the topic.

Use Role Play to Kickstart Topic Exploration

To explore a new topic, try, "Act as an experienced environmental researcher. Talk to me about the latest trends in sustainable packaging."

This approach helps you access in-depth knowledge and dive into a topic from a specific perspective, making it easier to get focused, relevant information right from the start.

Follow the Who/What/How Framework for Clear Prompts

Make sure your prompt clearly defines who AI is, what the task is, and how you want it done. This framework ensures you're giving AI the full context it needs to provide accurate and high-quality responses.

Remember, You Can Change Anything Anytime

Don't hesitate to make multiple changes or tweaks to your prompts. Unlike human collaborators, AI won't get tired or frustrated with repeated edits. Test different approaches until you get exactly what you need.

Rephrase or Break Down Tasks If AI Doesn't Get It

If AI isn't understanding your request, try rephrasing it or breaking it into smaller, step-by-step tasks. This can clarify what you're asking and help AI deliver more accurate responses.

Save and Share the Thread URL for Easy Access Later

To revisit important threads, save or share the URL link. If you need AI to analyze a past conversation, consider copying key sections into a new thread rather than uploading the entire chat—AI performs better with shorter, focused text.

Start a New Thread If AI's Responses Decline in Quality

If responses suddenly become less helpful, your thread may have become too long. Copy and paste relevant context into a new thread to refresh the conversation and get back on track with high-quality outputs.

Ask AI to "Push Back" or Challenge Your Ideas

Prompt with, "Challenge my ideas or suggest areas where I might be overlooking potential issues."

AI can serve as a valuable sounding board. Encourage it to question your assumptions or offer counterpoints, helping you refine ideas or spot potential flaws.

Have AI Ask You Questions to Complete a Task

Instead of saying, "Create a social media plan for my business," try, "Ask me any questions you need to create an effective social media plan for my business." AI might respond with questions like, "What are your main goals for social media?" or "Who is your target audience?"

This approach prompts AI to gather the details it needs for a specific, relevant output. It's especially helpful for complex tasks, where gathering information upfront leads to better, more tailored results.

Made in United States
Cleveland, OH
05 November 2026

47670537R00160